Fictions in Science

Routledge Studies in the Philosophy of Science

Fictions in Science

Philosophical Essays on Modeling and Idealization

Edited by Mauricio Suárez

Routledge
Taylor & Francis Group
New York   London

First published 2009
by Routledge
270 Madison Ave, New York, NY 10016

Simultaneously published in the UK
by Routledge
2 Park Square, Milton Park, Abingdon, Oxon OX14 4RN

Routledge is an imprint of the Taylor & Francis Group, an informa business

© 2009 Taylor & Francis

Typeset in Sabon by IBT Global.
Printed and bound in the United States of America on acid-free paper by IBT Global.

The essay in chapter 1 first appeared as Fine, A. (1993), "Fictionalism," *Midwest Studies in Philosophy*, Vol. 18, pp. 1–18, and is reprinted here courtesy of Blackwells' publishers.

Library of Congress Cataloging in Publication Data

Fictions in science : philosophical essays on modeling and idealization / edited by Mauricio Suárez.
    p. cm. — (Routledge studies in the philosophy of science ; 4)
    Includes bibliographical references and index.
    ISBN 978-0-415-99035-6
    1. Science—Philosophy.   2. Science—Methodology.   I. Suárez, Mauricio.
    Q175.F474 2009
    501—dc22
    2008018425

ISBN10: 0-415-99035-1 (hbk)
ISBN10: 0-203-89010-8 (ebk)

ISBN13: 978-0-415-99035-6 (hbk)
ISBN13: 978-0-203-89010-3 (ebk)

Contents

Acknowledgments

The idea for this volume came to me during a small workshop that took place in February of 2006 at Complutense University in Madrid. Some of the papers collected here were delivered then in an embryonic form. I would like to thank the participants at this workshop, as well as those contributors who joined the project at a later stage, for their enthusiasm and encouragement throughout. Arthur Fine in particular inspired my interest on fictionalism over a decade ago at Northwestern University, and his support has been constant and unfailing since. The award of a Leverhulme Trust Fellowship during 2001–2 allowed me to start working on this project. Later on, the institutional support from the Rector's Office at Complutense University, and in particular the Vicerectors of Research (Carlos Andradas) and International Relations (Lucila González Pazos), was decisive. Financial support is also acknowledged from the Spanish Ministry of Education and Science (research projects HUM2005–01787-C01–03 and HUM2005–0369). In addition I would like to thank all those people who chaired sessions and contributed to the discussions during the workshop. Thanks also to the members of my research team who helped with logistics and the organization. Albert Solé in particular provided editorial help in the final stages of putting the volume together. Erica Wetter from Routledge was an enthusiastic and supportive editor from the start. On a more personal note, Samuel Suárez taught me much about hobbyhorses, gameplaying, and the imagination in general. This effort is dedicated to him.

Madrid, December 2007

Part I

Introduction

1   Fictions in Scientific Practice

Mauricio Suárez

1.1 FICTIONALISM IN THE PHILOSOPHY OF SCIENCE

This volume collects thirteen essays by prominent contemporary philosophers of science on the role that fictions and fiction-making play in the practice of theorizing and model-building in science. The topic itself is not new in philosophy of science, because fictionalism has a long history in the discipline that goes back at least to the writings of Hans Vaihinger in the early years of the 20th century. Although Vaihinger's work was very popular in his own time, it fell into a kind of oblivion for years afterward, coinciding with the rise of logical positivism. The topic made a return to the philosophy of science agenda about 15 years ago, partly in the wake of debates about Bas Van Fraassen's constructive empiricism, which can be construed as a kind of fictionalism about theoretical entities,[1] and partly as a result of the increased attention paid by philosophers to models and modeling in the sciences. However, within philosophy of science the discussion over fictions has so far been rather fragmentary, with references to fictionalism typically playing some minor rhetorical role in the realism–antirealism debate, and appeals to fictions and fictive entities appearing occasionally in diverse case studies devoted to modeling. This might be contrasted with the very extensive treatment of fictions in the philosophy of language and aesthetics literature over at least the last two decades.[2]

The present volume is intended to help to redress the balance. It is the first volume that treats fictions from a philosophy of science perspective.[3] It is also the first collection of papers entirely devoted to the topic of fictions and fictionalizing in scientific practice.[4] It represents the most recent and up-to-date thinking on the topic over a range of scientific disciplines in contemporary philosophy of science. The authors reflect upon the role and function of fictionalizing in the process of building models for natural and social systems. Hence the emphasis is markedly on case studies rather than general philosophical theory, because even the most general philosophical essays refer extensively to case studies from diverse sciences. Thus although the focus is often on the physical sciences, there are also essays on biology and economics, and references to case studies in other areas of science appear throughout the essays.

1.2 FINE'S VAIHINGER

Although philosophers of science have paid scant attention to fictions, there have been some exceptions. In 1993 Arthur Fine published a seminal article entitled "Fictionalism" where he attempted to bring Vaihinger's philosophy of 'as if' back to the forefront of philosophical debate. Besides promoting a reevaluation of Vaihinger's work, this essay for the first time makes an explicit connection between fictionalism and the modeling literature in the philosophy of science of the last two decades or so. Fine's essay is already cited in anthologies as a classic "revival" piece, and it is the standard contemporary reference on Vaihinger among both philosophers of science and metaphysicians. And because most of the chapters in this volume refer to this essay, often quoting it as a source of inspiration, it is here reprinted as the starting chapter of the book. It appropriately sets the tone and the terms for most of the other essays in the book.

Fine's paper is not a piece of scholarly history but an interested commentary on a historical figure and its relevance for the contemporary debate. In other words, Fine's *Finehinger* is bound to differ from the original Vaihinger, and Fine himself acknowledges some unhistorical elements (Fine, this volume, section 8). The account is arguably anachronistic in the following two senses.[5] An earlier appeal to Vaihinger appears in Bas Van Fraassen's celebrated book *The Scientific Image* (Van Fraassen, 1980, p. 35–36), where Vaihinger's philosophy is cited in support of constructive empiricism's agnosticism regarding theoretical entities. More specifically, Vaihinger's fictionalism is invoked there as ammunition against the verificationist's commitment to the full theoretical equivalence of empirically equivalent theories (roughly, on this account, if two putatively different theories have exactly the same empirical consequences then they really are just the same theory). Van Fraassen argues that by contrast the fictionalist can claim that a theory that postulates a fictional unobservable entity is distinct from its empirically equivalent non-entity-postulating rival theories—even though none of these theories commits in fact to the reality of the entity. And this is just the constructive empiricist's response to the verificationist challenge. Thus, although Van Fraassen does not make the mistake of explicitly pinning the constructive empiricist commitments on Vaihinger, the reader of these passages might be led to suppose a strong association between Vaihinger's fictionalism and present-day antirealism of the constructive empiricist variety.

The association finds echoes in a few passages in Fine's piece, but it is definitely not Vaihinger's—who could not possibly have anticipated logical positivism's verificationism, Van Fraassen's response to it, or the main arguments in the contemporary realism–antirealism debate. Vaihinger himself was not committed to a fundamental epistemological difference between our knowledge of the observable world and that of the unobservable world. It is even questionable whether he acknowledged the antecedent distinction between observable and unobservable entities or domains of the world.

Another anachronism is arguably Fine's close association of Vaihinger to Neurath. This nowadays looks out of place with Vaihinger's pronounced neo-Kantian *transcendentalism*.[6] The association probably responds to the attempts in the early 1990s to re-evaluate logical positivism by emphasizing its "sociological" strand, and the then comparatively neglected figure of Neurath. Fine is here trying to distinguish Vaihinger's project from Carnap's and Reichenbach's brand of "logicism," and the most straightforward way to do this in the climate of the early 1990s is to associate Vaihinger with the rising counterbalancing figure of Neurath. Arguably, the real Vaihinger is a little less committed to the particularism defended by Fine's Natural Ontological Attitude (NOA). After all, Vaihinger promotes a theory of scientific reasoning, and advances what he takes to be a general theory of the logic of scientific fictions. Both assumptions seem estranged from Fine's NOA, and would have been alien to Neurath. Crucially no anticipation is to be found in Vaihinger of Neurath's most valuable and long-lasting contribution to the theory of knowledge, namely: his holistic epistemology regarding empirical knowledge.

Regardless of any anachronisms, Fine's article has rightly set the agenda for the discussion of fictions in contemporary philosophy of science, and most of the essays in the volume refer to it in such terms.[7] In particular, Fine lays out the connection between fictions and the contemporary philosophical literature on modeling practice more carefully and helpfully than Vaihinger would have been able to do. For Fine the recent literature on idealization and abstraction in science backs up Vaihinger's views, whereas all cases of scientific model-building are instances of Vaihinger's fictionalism at work: "Preeminently the industry devoted to modeling natural phenomena, in every area of science, involves fictions in Vaihinger's sense" (Fine, this volume, p. 34). The essays in this volume address this claim in a variety of ways, sometimes by filling in detail through case studies, sometimes by adding philosophical flesh and argument, and yet at other times by taking issue with the claim or qualifying it to some extent.

1.3 A MAP OF THE BOOK

Part II of the book ("The Nature of Fictions in Science") contains two essays besides Fine's devoted to the nature of fictions in science. Joseph Rouse (Chapter 3) defends the radical thesis that fictions and fictionalizing are not only present in modeling, simulation, supposition, and thought-experimentation. These are activities that we may call "representational" in a broad sense, and may seem prima facie distinct from most of the activities carried out in experimental science, which typically involve actual causal interventions on experimental systems in real laboratories. Fictions and fictionalizing have typically been thought to be confined to the realm of the "possible" within representational science. Rouse defends the view that fictionalizing also plays a role in the constitution of objects and procedures actually employed in the laboratory sciences. He appeals to case studies from genetics and thermometry in order to back up the claim that hypothetical entities and fictional assumptions are embodied in the material construction of these objects. (The same view is defended by Ankeny in Chapter 12 in relation to model organisms in biology.) The point of such entities and assumptions, according to Rouse, is to introduce alternative concepts, thus opening up a new space of reasons for the justification of scientific belief. Thus fiction-making in science can not be reduced, argues Rouse, to an exercise of expediency whereby false assumptions are employed for inferential or predictive gain. Fictional assumptions are not simply false-but-useful; they actually define both what is false and what is useful. This is a radical new proposal regarding the nature of fiction in science, going beyond the claims by most other contributors to the volume, but very much deserving to be explored.

In addition, Rouse's emphasis on activity and practice ahead of ontology is nicely in line with the rest of the essays in the volume. The contributors to this volume are neither distracted nor particularly concerned with questions regarding *ontology*. The purpose of the book is instead to focus on relevant aspects of the *activity* of model building. It might seem at first sight that Chapter 4 is an exception to this rule, because Barberousse and Ludwig's aim is to fill much-needed detail into the claim—a slogan, really—that scientific models *are* fictions. This *identity claim* about the ontology of models has been made before—for example, by Cartwright (1983, pp. 151–162)—but Barberousse and Ludwig point out that both sides of the identity need to be explicated to take it beyond a mere slogan.

Moreover, the slogan fails to capture the claims defended by the contributors to this book, which are fundamentally distinct. The essays here collected focus on the important role that fiction-making plays within the activity of modeling in science. But they remain on the whole silent on the question "what is a model?" which they tend to view as an issue of secondary importance.[8] However Barberousse and Ludwig's approach to this issue is of a piece with the book's outlook, because they claim that the slogan "models are fictions" is best filled in precisely by putting activity center stage, as opposed to structure or nature, as an account of both models and fictions. In particular they follow Currie (1990) and Walton (1990) in understanding fiction as the activity of imagining and role-playing, as opposed to any putative referential relations between "representans" and presumed "representanda." They then go on to use this view in order to provide a typology of models in terms of their functions (prospective, bridge, and what they call "anti-Duhemian" models).

In keeping with the book's outlook, part III ("The Explanatory Power of Fictions") swiftly moves away from issues related to the nature of fiction toward a discussion of one of the main and most controversial roles that fictional assumptions may play in modeling practice. Recent philosophy of science has emphasized the importance of explanatory virtues in the assessment and evaluation of theories and models alike. A widespread assumption is that false theories that are known to be false may never have explanatory power. Scientific realism backs this up by appeal to the idea that the explanatory power of a theory depends upon its closeness to truth. (So an inference to the most explanatory theory guarantees that the one closest to the truth is selected.[9]) The antirealist disagrees that this is a valid mode of inference (the most explanatory theory need not be closer to the truth), but still finds it difficult to articulate a notion of "explanation" that will fit this bill. So the question "do fictional assumptions have explanatory power?" goes to the heart of the realism–antirealism dispute. The three essays collected in this part of the volume answer this question differently.

First, Catherine Elgin (Chapter 5) argues for a move away from the notion of explanation and toward a more general notion of scientific understanding. Her view is that representation in science is best understood along the lines of Goodman's notion of "exemplification" of properties. Roughly, a model A exemplifies a property B if and only if A is denotative of B and moreover A possesses the property B. (So a concrete model of a building exemplifies fragility if it denotes fragility and *is* fragile.) The advantage of this account, according to Elgin, is that representation does not require actual denotation, but only that the 'representans' be a "denoting symbol or system." In other words, actual successful reference is less important for representation than the *function* of denoting. Because the representational character of models does not depend on successful denotation, a model can be representational just in virtue of purporting to denote, regardless of its actual success in doing so. This allows Elgin to develop an extremely interesting and novel account of explanation, which she refers to as "understanding as exemplification." On this account, fictions and the fictional assumptions made within modeling practice can play a fully explanatory role. Elgin then goes on to argue that there is nothing suspiciously subjective or arbitrary about this form of explanation.

In Chapter 6, Alisa Bokulich takes the issue of the explanatory power of fictions head on, and argues that fictions can definitely be used for explanatory purposes. Her argument proceeds by looking at the details of a case study from contemporary physics, namely, the appeal to fictional classical trajectories in explaining anomalous quantum spectra in the so-called "closed orbit theory" of the atomic nucleus. Bokulich displays textual evidence that scientists find such classical trajectories explanatory in spite of their being known to be fictional. She then puts these explanatory practices on the table as benchmarks for different theories of explanation, and argues that neither Hempel's covering law account nor Salmon's causal-mechanical account can provide us with a similar philosophical understanding. Instead Bokulich shows that the practice of using fictions explanatorily points toward a further form of explanation, distinct from the previous ones, namely, what she calls "model explanation." In particular, she claims that the explanatory power displayed in her case study can be understood as a subspecies of this form of model explanation, namely, structural model explanation, where the model's structure is the key to the model's explanatory power.

In Chapter 7, Margaret Morrison takes up a famous case study from the history of 19th-century physics, Maxwell's model of the ether. This has of course come down in history as one of the most striking cases of fictions at work in science. Employed for all sorts of explanatory, predictive, and heuristic tasks, the mechanical models of the ether were self-conscious fictions for Maxwell and other Maxwellians throughout the 19th century. Morrison asks the question: How can this fictional assumption play a role in delivering accurate information regarding electromagnetic phenomena? She argues that these models play out their own story by constraining the values of certain important parameters. In addition, Morrison claims that in order to go beyond mere quantitative connections and understand how explanation works, we need to look into idealization and abstraction as separate from fiction. Morrison takes up a different case from genetics, the Hardy–Weinberg law, in order to illustrate these alternative possibilities. So for Morrison it is not the case that every unrealistic assumption employed in science is fictional. This seriously qualifies Vaihinger's original claim for the "ubiquity of fictions in science," restricting its domain of applicability. On the other hand, Morrison agrees that fictions do serve the maxim of expediency—in both reasoning and inference. She does argue, however, that there is no one unique method whereby they achieve such aim in practice.

All three essays in part IV of the book devoted to physics ("Fictions in the Physical Sciences") assume such pluralism regarding the methods whereby fictional assumptions maximize expediency. First, in Chapter 8, Carsten Held focuses on the introduction of posits into theories, and the status of their referential relations. By means of a case study in quantum field theory, Held shows that this status may well change, and often does so throughout the history of the development of a theory. Thus a theory's posit that nowadays has stable reference need not have had it when first introduced. Held is here providing a very nice illustration of Vaihinger's claim that fictions, which are typically not up for empirical 'grabs,' often turn into testable hypotheses—and although sometimes such hypotheses turn out to be (illustrious) mistakes, as in the case of the ether, occasionally they "make it good," successfully achieving stability of reference.

In Chapter 9, I illustrate some of Vaihinger's fundamental distinctions by means of four case studies, two celebrated cases from 19th-century physics (Maxwell's ether and Thomson's atom), and a further two from contemporary physics (models of stellar structure in astrophysics and the quantum model of measurement). The aim is to show that fictions are not just discarded suppositions of previous failed science, but they play a role in established and accepted science too. I argue that the quantum measurement models provide a striking illustration of the *full fictions* involving self-contradiction that Vaihinger used to talk about, and which most commentators had previously thought to be impossible. Building on Vaihinger's and Fine's claims (shared by Morrison, this volume) I defend the expediency of fictions in inference and reasoning. I go further in arguing that the maxim of expediency in inference (together with a requirement on the empirical testability of consequences when conjoined with further hypotheses) distinguishes appropriately the function of the fictions one finds in science from fictions elsewhere, particularly in the arts. I then show that the inferential conception of representation appropriately accounts for these distinctions, and in particular explains naturally why the fictions one finds in science are governed by the maxim of expedient inference.

In Chapter 10, Eric Winsberg takes up the most pressing objection to Vaihinger's "ubiquity of fictions" claim. Like Morrison, Winsberg argues that of course science is full of idealizing assumptions that are strictly speaking false, and often known to be false by those scientists who employ them. Idealized frictionless planes and penduli abound in scientific models since Galileo's time. But this does not mean, argues Winsberg, that these assumptions are fictional. To be fictional an assumption need not be false (although it may happen to be false), but rather *unconcerned* with the truth. This is an important insight shared by a number of essays in the volume (for more on this topic see section 1.5 in this introduction). But Winsberg goes on to show that it is an insight that does not compromise the fictionalist outlook. Although fictions may not be as ubiquitous in science as Vaihinger and Fine supposed, they are nonetheless often present. Winsberg describes a couple of case studies in support, from nanomechanics and computational fluid dynamics. In both cases fictions are invoked in order to bring or patch together at least two indispensable but otherwise incompatible accounts of the phenomena. The role of fictions is to avoid straight contradiction in such cases, allowing for a variety of empirical predictions to be carried out.

The fourth part of the book ("Fictions in the Special Sciences") takes us away from physics and into other disciplines, which so far have had limited exposure in the book. In Chapter 11, Rachel Ankeny advances the case for fictions in the biological sciences. She first argues that fictional assumptions can generally be mobilized in order to gain real knowledge of the world. A novel and important aspect of Ankeny's claims is that the fictional character of an assumption is not an absolute but a relative matter—it depends on what alternative assumptions it is contrasted with. Ankeny then provides a review of the literature on model organisms, in order to show that different kinds of model organisms in biology display all the characteristics of fictitious entities to some degree. In particular she argues that novel synthetic organisms display a particularly striking form of fictitiousness: These are artificially constructed objects built and created in order to mimic certain salient aspects of real organisms. In other words, these systems provide an illustration of the thesis that we saw Rouse defending in Chapter 3, namely, that fictional assumptions can sometimes be embodied in real systems and objects.

Chapter 12 raises similar issues for economics. Tarja Knuuttila focuses in particular on the much-debated issue regarding the realisticness of the assumptions in economic models. The issue here is whether such assumptions may be used locally in order to generate models that are globally realistic, as has been argued for instance by Mäki (1994). Knuuttila raises a number of problems with this view in economics, and goes on to argue that a better way to approach the issue takes models to be autonomous and independent entities that can only represent indirectly, in agreement with a pragmatist and inferential conception of representation. Models are then seen to have some of the properties of fiction (while avoiding a wholesale identification of models with fictions). Knuuttila then faces up the greatest challenge for this view, namely: How can we learn anything about the world from considering such autonomous and independent entities? Her answer is of a piece with many of the book's other essays, and assumes that further assumptions will bring out the real-world, or at least empirical, consequences of the model.

The last part of the book ("Fictions and Realism") looks critically at the epistemological consequences of the book's overall fictionalist outlook, particularly as regards the debate over whether science aims at truth. These two essays, by Paul Teller and by Ron Giere, are among the most critical. They both argue that we should not generally identify scientific models with fictions, and that we should not therefore be led to believe that science does not aim at truth, or "truth enough." Although there are fictional assumptions made in the construction of many models, this does not mean that the models themselves are wholly unconcerned with the truth. Rather, the false assumptions made along the way may even help the model get overall closer to the truth. Like Morrison and Winsberg, Teller and Giere operate with a limited notion of fiction, and attempt to show that idealization and abstraction trade in falsity or omission too, but in ways that are innocuous to the realist. The contrast is greatest at this point with Fine, Rouse, Bokulich, Suárez, Ankeny, and Knuuttila—because all these authors tend to follow Vaihinger in looking at fiction and make-believe as a more general category that encompasses crucial aspects of the others. But as a matter of fact the disagreement is less than it might seem. Teller and Giere are particularly critical of the thesis that models are to be identified with fictions. But it has already been noted that the essays in this volume are not actually out to defend this thesis—some of the authors even make it explicit that they find the question about the ontology of models to be irrelevant, or uninteresting. And when it comes to the activity of model-building, Teller's and Giere's essays, like all the others in the book, are able to unearth a variety of fictional assumptions made by scientists in both theorizing and modeling, thus contributing to the book's general outlook.

1.4 FICTIONALIZING: PRACTICE AHEAD OF ONTOLOGY

The book's focus is unashamedly on the practice of scientific modeling, rather than the abstract nature of models. This means that there is a preference for considering the nature and structure of models, if anything, within the context of the activity of model-building. And rather than considering the metaphysics of the relations of representational sources and targets, the essays focus on the kinds of activities that representations allow scientists to carry out. In other words, in terms of the usual dichotomies, the book emphasizes process over product, procedure over outcome, function over essence, praxis over axiom, activity over ontology.

Similarly, when it comes to fictions, questions about the nature, ontology, and essence of fictitious entities are always framed within the context of the activity of fictionalizing, make-believe, and fiction-making. There exists a huge philosophical literature on the metaphysics and semantics of fiction, with a strong focus on the ontology of fictional entities. The classic positions range from Meinong's notoriously reifying fictional entities to Quine's wholesale dispensing with any abstract entities altogether.[10] This literature links the question of the nature of fiction to the coupled issues of the ontology of fictional entities, and the semantics of the terms and names that our natural languages make available for the putative denotation of such entities. In other words, the issue of the nature of fiction has traditionally been linked to the issue of linguistic reference.

This book takes an alternative approach, which it believes to be more productive for philosophers of science. In particular, the contributors express no anxieties at the lack of a compelling *semantic theory* of fictional terms. Instead, the essays in the book focus on the myriad ways in which fictions are put to use in science, and the role that the activity of fiction-making plays in scientific modeling in particular. Thus although the issue of the nature of fiction is not avoided, it is not answered in any of the ways that would be typical of metaphysical inquiry. Questions regarding the nature of fiction are rather addressed by studying the cognitive functions that fiction plays in scientific inquiry, and in particular in the practice of model-building.

1.5 FICTIONS: TRUTH-CONDITIONAL
VERSUS FUNCTIONAL ANALYSES

There is consensus throughout the book that fictions and fictional assumptions play a key role in modeling practice (with some qualified dissent expressed in the essays by Paul Teller and Ronald Giere in the last part of the book). This is achieved in spite of an array of different views by the contributors on the defining conditions for fictions. Some of the essays (Bokulich, Morrison, Teller, Giere) tend to follow Vaihinger in viewing fictions in modeling as knowingly false assumptions about the systems modeled. A fiction on this account is a description that involves falsehood (or a representation that involves inaccuracy), and one that is known to involve it by the scientists who nonetheless employ it.

We may refer to this as the *truth-conditional account of fiction*, because for an assumption to function as fiction in a model it is required that its users understand the truth conditions of the assumption and be able to assess that it is false. (Or, for a representation, understand its application conditions, and assess it to be inaccurate.) On this account the assumption of the mechanical ether is a fiction for Maxwell, who always assumed his models of the ether to be false in terms of the real causes of the electromagnetic phenomena. But the very same assumption is not a fiction for Trouton or any of the other 19th-century physicists who believed in the reality of the mechanical ether.

However, other contributors (notably Rouse, Elgin, Winsberg) follow a distinct lead in Vaihinger's thought, and defend the view that fictions ought to be characterized functionally by their role in inquiry, rather than truth-conditionally. The character of fiction depends on this view upon the nature of the cognitive processes and functions that fictions allow scientists to perform; nothing substantial hinges on whether the fictions, or fictional assumptions within a model, are true or false. In fact on this, *the functional account of fiction*, the key to the nature of fiction in science lies precisely in the irrelevance of its truth value for its cognitive function. When the function within a model of an assumption about x, or a representation of x, is entirely independent of its truth value, or correctness, we are then in the presence of a fiction. The key to fictionalizing on this account is the free exercise of the imagination that comes with such a release from a concern with truth and truth-conditions. The scientists who employ fictions are required neither to understand their truth conditions nor to assess its truth value. On this account the mechanical models of the ether are fictional for Maxwell, Trouton, and all other 19th-century physicists, because its main role is to bring into agreement frameworks that would otherwise remain separate, and not to get at the truth. Winsberg's *silogens* are a remarkably clear instance of this view.

Finally there are mixed cases. Fine's Vaihinger endorses both the truth-conditional and the functional accounts—that is, fictions are in part defined by their being known to be false, and in part by their non-truth-driven function in inquiry. Both Held and Ankeny defend both accounts simultaneously too. Similarly, Suárez adopts both accounts for fictions in general, but employs only the functional account to distinguish specifically scientific fictions. Similarly, Knuuttila explicitly employs the functional account, but takes fictions to provide arguments against realism—by assuming that the truth-conditional account results as a by-product.

1.6. FICTIONALIZING, IDEALIZING, ABSTRACTING: EPISTEMOLOGICAL DISTINCTIONS

We may in general distinguish two different kinds of falsehood for any assumption (or two kinds of inaccuracy for any representation). An assumption about some entity x may be false (or, a representation of x may be inaccurate) because there is no such entity in reality. We may refer to this as a *fictional* assumption (or, a *fictional* representation). On the other hand the same assumption may be false because the entity x, albeit real, is incorrectly described, perhaps because the properties that are ascribed to it in the assumption or representation are not the properties that x actually has. We may refer to this as a *fictive* assumption (or *fictive* representation). Both types of assumption or representation involve falsehood or inaccuracy, but the kind of knowledge that is required to establish their error is very different. (See chapter 9 for further elucidation of the distinction.) For 19th-century physicists the assumptions of the mechanical models of the ether are false in the latter but not the former sense: They are fictive but not fictional. By contrast from our own, post-Einsteinian, point of view all theories of the ether are fictional, because they attempt to describe nonexistent entities. The truth-conditional account of fiction is consistent with both but requires a very different assessment of the truth conditions for both types of assumptions. In particular, the falsity or inaccuracy of fictional assumptions and representations requires the knowledge that the entity in question does not actually exist.

The contributors to the volume differ in the extent to which they are prepared to accept fictive representation as a kind of fiction. Roughly for Fine, Barberousse and Ludwig, Bokulich, Held, Suárez, Ankeny, and Knuuttila, fictive assumptions and representations *are* fictions. We may say that these authors defend a thorough or "wide" fictionalism. By contrast, Morrison, Winsberg, Teller, and Giere think that what I have here called fictive representation is simply a case of idealization or abstraction. They instead choose to reserve the name "fiction" for fictional representations and assumptions only. Thus we may say that these contributors defend "narrow" fictionalism. *Wide fictionalism* takes it that idealization and abstraction are subspecies of fiction, and consequently that most modeling practice involves fiction and fictionalizing most of the time. *Narrow fictionalism* by contrast takes it that fictions are not idealizations or abstractions, and thus fictionalizing plays a less ubiquitous role in modeling. Perhaps not surprisingly, the promoters of *wide fictionalism* tend toward instrumentalism in the epistemology of science, whereas the defenders of *narrow fictionalism* are friendlier toward scientific realism.

The most pressing task for both *wide* and *narrow* fictionalism is to provide an account of the cognitive value of fictional assumptions within modeling practice. There is agreement throughout the book that in order to be cogent and cognitively useful the fictional world depicted by the model must incorporate plenty of assumptions derived from our experience. These further background assumptions will serve to extract consequences about the systems modeled from the fictional world depicted by the model. The instrumentalist will tend to think that the relevant experience is of the empirical kind; for the scientific realist this experience is of the real world. Similarly, the consequences extracted from the model will be about the empirical world for the instrumentalist, and about the real world in general for the scientific realist.

Those contributors who tend toward instrumentalism (particularly Fine, Suárez, and Knuuttila), also tend to think that *wide fictionalism* provides arguments for antirealism. Other contributors who promote *narrow fictionalism* (Morrison, Giere, and to some extent Teller) do not find that the presence of fictional elements is a threat to the scientific realist. On the contrary, they would argue that these elements of creativity, game playing, and imagination can help advance the realist cause. Hence the debate over the epistemology of science is not foreclosed once it is accepted that fictions play a role in scientific activity and modeling practice. The realism–antirealism debate instead takes on a new life, because both realists and antirealists must now bring their favorite arguments to bear on the matter.

Giere's concluding essay expresses with great clarity an objection against *wide fictionalism* that he takes to be fundamental. Giere claims that this all-reaching form of fictionalism poses a threat to scientific rationality, because it obliterates the distinction between science and other, myth-based systems of belief where fictions play a more ostentatiously prominent and ubiquitous role. However, the defender of *wide fictionalism* would deny this. He or she would point out that this is not the right 'joint' at which to 'carve' the undeniable differences between science and these alternative belief systems. Thus Giere's conjuring of the creationism debate, and the fictions of religious belief, is a timely warning and an important reminder that there *must be* differences between the roles of fiction and make-believe in science and elsewhere. It serves to remind us that there is important work for philosophers to do in carefully distinguishing such roles. But instilling a blanket fear of fictions in science would just be unproductive toward this goal. Fortunately, this volume ought to constitute good therapy against "fiction-panic"—for realists and antirealists alike. The book embraces the fictions that lie at the heart of science without complexes or fear. All the essays (including Giere's) make a sustained effort to bring out what is distinctly scientific about the use of fictions in science, in contrast to literary or artistic fiction. The authors of these essays avoid being swayed by mythological accounts of science—inspired by legend rather than fact—that presume at the outset that there is no room for game-playing or fictionalizing within the "grown-up" rationality of science.[11] Instead they come to terms with the functions and roles of fiction and make-believe in *actual* scientific practice. And, as scholars devoted to a greater understanding of our actual science and rationality, should we not prefer the actual fictions of our real science to the virtual hard facts of some mythical and inexistent pseudoscience? These essays provide a compelling picture of the myriad ways in which scientists' imagination, game-playing, and conceptual creativity serve to advance the goals of our scientific objectivity, and to genuinely promote the growth of our scientific knowledge.

NOTES

1. See, e.g., Rosen (1994, 2005).
2. Some key classic texts include Lewis (1983), Currie (1990), and Walton (1990).
3. Kalderon (2005) is a nice recent collection—but it is mainly devoted to issues of ontology, and to comparisons with cognate positions in areas traditionally thought to be friendly to fictionalism anyway, such as moral expressivism and error theory in ethics. The only essay in that collection that addresses the sciences is Rosen (2005), but it is not focused on contemporary science, but mainly on the history of astronomy—again a discipline traditionally thought to be relatively friendly to fictionalism.
4. Mäki (2002) is mainly about the realism issue in economics. (See Chapter 12 in the present volume for a critical assessment of Uskali Mäki's own work on related themes.)
5. Fine argues that Vaihinger's work needs to be placed in the context of his own time, and similarly Fine's own piece needs to be placed in the context of the ongoing epistemological debates of the early 1990s—which is what I try to do in the main text earlier.
6. See particularly the long section in part 3 entitled "Kant's Use of the 'as if' Method" (Vaihinger, 1911, pp. 271–318). Thanks to Thomas Uebel for urging me to take account of Vaihinger's neo-Kantianism. Neurath's preference for the language of models over theories might certainly point to an affinity with contemporary philosophy of science's focus on modeling—see, e.g., Neurath (1983).
7. Including those essays by Rouse (Chap. 3), Morrison (Chap. 7), Suárez (Chap. 9), Winsberg (Chap. 10), Teller (Chap. 13) and Giere (Chap. 14) that in different ways take issue with some of Fine's and Vaihinger's claims!
8. Only Ron Giere (Chapter 14) comments on the slogan—but finds it defective. See section 1.4 in this Introduction for further justification for the book's outlook.
9. One locus classicus for a defense of inference to the best explanation is Peter Lipton (1991).
10. For a review of some of the relevant positions, see Sainsbury (2005).
11. 'Legend' is Philip Kitcher's apt name for the account of scientific method as a superior and streamlined form of rationality that inevitably leads toward the discovery of truth, and the ultimate avoidance of fiction and error (Kitcher, 1993). Kitcher has now abandoned *legend* in favor of a pragmatist-inspired account of science that is prima facie compatible with the fictionalist outlook promoted in this book.

Part II

The Nature of Fictions in Science

2   Fictionalism

Arthur Fine[1]

One can [say] that the atomic system behaves "in a certain relation, 'As If . . .'" and "in a certain relation, 'As If . . . ,'" but that is, so to speak, only a legalistic contrivance which cannot be turned into clear thinking. (letter from E. Schrödinger to N. Bohr, October 23, 1926)

1. INTRODUCTION

Over the last few years the realism–antirealism debate has produced a few arguments (see Fine, 1986) and a somewhat larger number of epithets. The residual realist groups, while divided among themselves into competing factions, are still set against views generically labeled antirealist. These antirealist views are (sometimes) differentiated as being empiricist, constructivist, instrumentalist, verificationist, or one of the three 'P's—phenomenalist, positivist, or pragmatist; and so on. When an especially derisive antirealist label is wanted, one can fall back on the term "fictionalist," coupled with a dismissive reference to Vaihinger and his ridiculous philosophy of "As If."[2] But what is "fictionalism" or the philosophy of "As If," and who was this Vaihinger?

2. SCHEFFLER'S FICTIONALISM

Within the standard literature in the philosophy of science (excluding the philosophy of mathematics[3]), Israel Scheffler (1963) contains perhaps the last extended discussion of a position called "fictionalism." In the context of exploring the problems and resources of a syntactical approach to scientific theories, especially the problems of the significance and justification of theoretical terms, Scheffler distinguishes between a Pragmatic Attitude (or pragmatism) and a Fictionalist Attitude (or fictionalism). Pragmatism takes the existence of effective and systematic functional relations as sufficient for rendering discourse significant, that is, capable of truth or falsity. Thus pragmatism, in Scheffler's terms, holds that there may be a language that is significant throughout and also capable of expressing all of science. Fictionalism, by contrast, employs some criterion of intuitive clarity as necessary for significant discourse. Where this criterion is not satisfied we have a "fiction."

One kind of fictionalism is *eliminative;* it seeks to eliminate and/or to replace the fictional part of scientific discourse. The eliminative project is to construct a language that is thoroughly significant and also capable of expressing all of science. The unreconstructed language of science will not do. Another kind of fictionalism is *instrumentalist.* Instrumentalism withholds the language of truth, evidence, and belief from the fictional part of science, but holds that considerations relating to interest and utility are nevertheless sufficient for retaining that fictional component. Thus instrumentalism would suppose that any language capable of expressing all of science has a fictional (i.e., nonsignificant) component. On this view there is not much difference between pragmatism and instrumental fictionalism except over the question of whether significant discourse has to be intuitively clear. What is more important is that they both would tolerate discourse that fails to meet the criterion of clarity, provided the discourse were suitably functional.

Scheffler's discussion of fictionalism is bound to the 1960s project of telling a philosophical story about the regimentation of science as a whole. Specifically, it is tied to that project's preoccupation with the syntax of the regimentation and the problem of meaning for so-called 'theoretical terms.' Thus Scheffler redrafts the concepts of pragmatism, instrumentalism, and fictionalism to suit his global, syntactic approach and his meaning-related needs. Those familiar with the American pragmatism of Peirce or James, or with the instrumentalism of Dewey, will find that Scheffler's terminology points at best to shadows of fragments of the originals. The same is true with respect to the fictionalism that was Hans Vaihinger's.

3. VAIHINGER AND LOGICAL POSITIVISM

We can begin to appreciate the importance of Vaihinger and his fictionalism for the philosophy of science by looking at the connection with the Vienna Circle that formed around Schlick, Carnap, and Neurath in the 1920s and the similar "Vienna Circle" of a decade earlier, also involving Neurath, along with the mathematician Hans Hahn and the physicist Philipp Frank. The term "logical positivism" became the catchword of the movement represented by these circles (and also the Berlin group around Reichenbach) following the article with that title by Blumberg and Feigl (1931). But that term did not originate in the manifestos of Neurath, nor in the writings of other dominant figures associated with these circles. Rather, as Philipp Frank notes (1949, p. 43), the first use of "logical positivism" comes from Hans Vaihinger's (1911) *The Philosophy of "As If."*[4]

Vaihinger described his philosophy using a number of different terms, including *critical positivism, idealistic positivism*, and also (p. 163) *logical positivism*. Here the modifier "logical" (as also "idealistic") refers to a logical or mental construct and the "positivism" insists on there being a suitable observational or experimental demonstration before one associates any reality with the construct. In describing his view as a kind of positivism, Vaihinger was trying to associate himself with an empiricist approach to positive, scientific knowledge and to disassociate his view from rationalism or Platonism—indeed, from any view that would presume some reality to correspond to whatever the mind logically constructs. Thus the logical positivism that Vaihinger introduced was an anti-metaphysical position of the same general kind as the logical positivism (or empiricism) usually associated with Vienna. Despite this similarity of philosophical orientation, however, Vaihinger is seldom cited by the Vienna positivists as a precursor, and never as an ally. To the contrary, as with today's realists, the writings of the logical positivists generally contain only the most curt and disparaging references to Vaihinger's central ideas.

Thus in Schlick's (1932) well-known reply to Max Planck's attack on Machian positivism, Schlick writes "if . . . Hans Vaihinger gave to his 'Philosophy of As If' the subtitle an 'idealistic positivism' it is but one of the contradictions from which this work suffers" (Schlick, 1932, p. 85). At the end of that essay Schlick says explicitly that the logical positivism to which he subscribes "is not a 'Theory of As If'" (Schlick, 1932, p. 107).[5] With these faultfinding remarks Schlick attempts to distance the Vienna school from Vaihinger, and misleadingly identifies Vaihinger's "idealistic positivism" and his 'As If' as a species of idealism, rather than as marking a positivist or empiricist attitude toward *ideas*. Similarly, Philipp Frank, who acknowledges Vaihinger's view as "the school of traditional philosophy that was nearest to [Vienna positivism] in spirit and in time" (Frank, 1949, p. 42), is still at pains to distinguish the two, which he does by denigrating Vaihinger as having a "complete lack of understanding" of what Frank regards as the important distinction between a coherent conceptual system and the operational definitions that connect it with the world of facts (Frank, 1949, pp. 42–43). One can hardly read these protestations of the significant differences between Vaihinger and the logical positivists without getting a strong feeling that, indeed, they protest too much.

4. VAIHINGER AND THE PHILOSOPHY OF "AS IF"

Hans Vaihinger was born in Nehren, Württemberg, in 1852. He received a religious and philosophical education at Tübingen and then did postgraduate work at Leipzig, where he also studied mathematics and natural science. He was Professor of Philosophy at Halle from 1884 until 1906, when he had to resign his position due to failing eyesight. He died at Halle in 1933. His *Philosophy of "As If"* first took shape as his dissertation of 1877. It was not published, however, until 1911. Vaihinger says that earlier he considered the time for it not ripe. The book is subtitled "A System of the Theoretical, Practical and Religious Fictions of Mankind" and the central topic of the whole work is an account of what Vaihinger calls "fictions" and "fictive judgments."

The book underwent extensive revisions and extensions in a very large number of editions. The sixth edition was translated into English by C. K. Ogden in 1924, with a second English edition in 1935, and several reprintings thereafter. In 1919 Vaihinger (and Raymund Schmidt) initiated the *Annalen der Philosophie*, a journal originally devoted entirely to the philosophy of "As If," and also committed to fostering interdisciplinary contributions, especially from the special sciences. The positivist Joseph Petzoldt joined as a third editor in 1927. By 1929, with Vaihinger blind and Schmidt effectively in control, the *Annalen* had fallen on hard times. After prolonged and delicate negotiations, the journal was taken over by Carnap and Reichenbach, to be reborn as *Erkenntnis*, the official organ of the logical positivist movement.[6] Earlier Vaihinger had founded *Kantstudien*, still today a leading journal for Kant scholarship. In addition to his "As If," Vaihinger wrote a well-known, two-volume commentary on Kant's first critique, where he developed the so-called "patchwork thesis" that distinguishes four layers relating to the subjective sources of knowledge in the Transcendental Deduction. He also wrote a work on Nietzsche and completed several other philosophical studies. Vaihinger is usually regarded as a neo-Kantian, although his reading of Kant was very idiosyncratic. For example, where Kant generally considers scientific principles as providing the possibility of objective knowledge (i.e., as constitutive), for Vaihinger in large measure (although not totally) scientific principles are fictions, functioning as regulative ideals. Overall, Vaihinger's work, in fact, shows a strong British influence—especially due to Berkeley on the philosophy of mathematics and Hume on impressions and the imagination. We shall see that, in many respects, Vaihinger is closer to American pragmatism than to the transcendental idealism of Kant. Indeed, Ralph B. Perry and other keepers of the pragmatic tradition identify Vaihinger, Williams James, and Poincaré as leading pragmatist thinkers of their era.[7]

Thus, despite its current eclipse, in his own time Vaihinger's fictionalism was widely known and, like the related work of Mach and Poincaré, it had its own strong following. Moreover, judging by the overall reaction, I am inclined to think that the impact of Vaihinger's work then was not unlike the impact of Thomas Kuhn's work in our time. In the 1960s and 1970s most philosophers of science reacted to Kuhn with strident criticism. The response of their logical positivist forbears to Vaihinger, as we have seen, displayed a similarly hostile tone. Today, however, notwithstanding that critical rejection, most English-speaking scientists, and commentators on science, are familiar with Kuhn's basic ideas about "paradigms" in science and freely employ his language. The same was true of German-speaking scientists and commentators with respect to Vaihinger's fictions and his language of "As If" up until the Second World War.[8] (Note, in the epigraph, Schrödinger's use of the "as if" idiom and his only slightly veiled reference to what Vaihinger called "legal fictions.") Except in discussions of legal philosophy,[9] however, Vaihinger did not survive the intellectual sea change that followed the war and restructured the philosophical canon. What I should like to do here is to review the main features of Vaihinger's work on fictions and then focus on a few central issues, showing their relevance to contemporary science and discussions in the philosophy of science.

5. FICTIONS

Vaihinger's general concern is with the role of fictional elements (or "fictions") in human thought and action. He begins with an elaborate classification of these elements and proceeds to illustrate their variety and use in virtually every field of human endeavor that involves any degree of reflective thought. Thus he illustrates the use of fictions in economics, political theory, biology, psychology, the natural sciences, mathematics, philosophy, and religion. Vaihinger acknowledges earlier treatment of fictions by Jeremy Bentham and he completes the survey by examining related conceptions that he finds in Kant, Forberg, F. A. Lange, and finally in Nietzsche. This survey is designed to achieve an effect. Vaihinger is trying to show that fictions are everywhere, that fictive thinking (like deductive and inductive thought) is a fundamental human faculty, and that—however dimly or partially—the importance of fictions has been recognized by the great thinkers. The extent to which Vaihinger's work was consumed by the philosophical public is evidence that, in good measure, he achieved the desired effect. Even his critics, like Morris R. Cohen in this country (see section 6 of this chapter), had to admit that Vaihinger had succeeded in getting people to realize the importance of fictions—reluctantly, even in science.

Vaihinger uses the term "fiction" loosely, sometimes for concepts and sometimes for propositions. He begins his account by distinguishing what he calls "real" or "genuine" fictions from what he calls "semi-fictions." Real fictions are characterized by three features: (1) They are in contradiction with reality, (2) they are in contradiction with themselves, or self-contradictory, and (3) they are generally understood to have these features when they are introduced. The semi-fictions satisfy (1) and (3) but not (2); that is, they are generally understood to be in contradiction with reality, but are not also self-contradictory. Following what one might call the Puritan principle,[10] Vaihinger also distinguishes between virtuous and vicious fictions, those that are scientific and those that are unscientific. The scientific fictions are an effective means to certain ends; they are useful and expedient. Where this utility is lacking the fictions are unscientific. An example of the unscientific kind would be the introduction of a so-called "dormative power."

The two primary examples of virtuous and real fictions with which Vaihinger begins are atoms and the Kantian "thing-in-itself." With respect to atoms his problem seems to be the difficulty of reconciling the conception that he takes from Cauchy, Ampère, and others, that atoms are centers without extension, with the idea that they are also the substantial bearer of forces. Vaihinger regards this as "a combination . . . with which no definite meaning can be connected" (p. 219) and so, presumably, as contradictory. Moreover, because he thinks that the idea of a vacuum is itself contradictory (following Leibniz here), the fact that there is supposed to be a vacuum between atoms further implicates the atom in contradictions as well. Still, Vaihinger recognizes the usefulness of the chemical atom, for example, in organizing chemical combinations by definite proportions; hence the expedient and scientific nature of the atomic fiction. I will pass over the "thing-in-itself" to give some examples of semi-fictions.[11] Vaihinger suggests that in 18th-century France the limitations and inaccuracies of Cartesian vortex theory were already known, at least to some, yet it remained a useful way to organize the motions of bodies. This may not be a very good example, because it is not clear that anyone ever understood how vortex theory was supposed to work. (So maybe this fiction is not scientific at all!) But then Vaihinger also calls attention to the status of Ptolemaic astronomy, the limitations of which he says were known by Arab scholars in the Middle Ages. Still, the Ptolemaic system was a useful (and so scientific) way to deal with the heavens—and so a scientific (semi-) fiction. Both of these examples are of what Vaihinger calls "heuristic" fictions, and they may help us understand what Vaihinger means when he says that a fiction is in contradiction with reality: namely, that it is in some measure not true to what it purports to refer. It is not so easy to understand what he has in mind, however, when he regards the genuine fictions as involving a self-contradiction. More on this shortly. First let me run through some of the fundamental distinctions that Vaihinger develops.

To begin with he classifies fictions into 10 primary kinds, with a few subcategories.

1. Abstractive
2. Of the Mean
3. Schematic
      Paradigmatic, Rhetorical, Utopian, Type
4. Analogical
5. Legal (Juristic)
6. Personificatory
      Nominal
7. Summational
8. Heuristic
9. Practical (Ethical)
10. Mathematical

Generally the name that he uses indicates the salient fictional feature. For example, abstractive fictions neglect important elements of reality. Thus we have Adam Smith's assumption that all human action is dictated by egoism, which looks at human action as if the sole driving force were egoism; or the treatment of extended bodies as if all their mass or gravity were concentrated at a point—and so the fictions of point masses and centers of gravity. Another abstractive fiction is what we would call the fiction of Robinson Crusoe worlds. This treats language as if it developed in worlds containing only a single individual. Vaihinger, however, holds to a social conception of language which makes this idea of a private language quite impossible of realization. Other kinds of fictions yield similar examples. These 10 categories are not disjoint. Thus a personificatory fiction involves an analogy to a person, as in God the father, but this is also an analogical fiction. There are countless other overlaps as well. Nor is it clear whether this list is intended to be complete; presumably not, because for Vaihinger even the Kantian categories are not to be regarded as fixed.

Vaihinger does suggest, however, that as we move down the list we are likely to move away from semi-fictions and toward the genuine thing. By the time we get to mathematics, with its various number systems, and the limiting geometrical concepts of points, lines, surfaces—not to mention the fluxions and differentials of the calculus—we have arrived at a really fictional realm. Echoing Berkeley (and Dewey), Vaihinger writes, "Mathematics, as a whole, constitutes the classical instance of an ingenious *instrument,* of a *mental expedient* for facilitating the operation of thought" (p. 57).

I have remarked that Vaihinger has a dynamic view of categories and classifications (as did later neo-Kantians like Cassirer, Reichenbach, and even Einstein). For example, at one point he says that the distinction between semi-fictions and genuine fictions is not stable in time. Often we learn that what we took for a contradiction with reality also involves a self-contradiction, so the semi-fictions may become real ones. Of greater importance for him is the distinction between hypotheses and semi-fictions,[12] for where we talk of Descartes's vortices or Ptolemaic astronomy (or Newton's laws, for that matter), it may seem that we are concerned with hypotheses and not with fictions at all. Vaihinger recognizes that when a scientific idea is first introduced we may not know whether it is a fiction or a hypothesis. We may begin by believing it to be the one and learn later that it is the other. What is the difference?

For Vaihinger, hypotheses are in principle *verifiable* by observation. We choose among hypotheses by selecting the *most probable.* In this way we *discover* which are true. By contrast, fictions are *justifiable* to the extent to which they prove themselves useful in life's activities; they are not verifiable. We select among fictions by choosing the most expedient with respect to certain ends. (He does not say that we maximize utility nor use Reichenbach's notion of *vindication,* but he might have.) Finally, fictions are the product of human invention; they are not discovered. In this connection he would have us contrast Darwin's hypothesis of descent with Goethe's schematic semi-fiction of an original animal archetype.

Showing his scientism, Vaihinger suggests that Goethe's fiction "prepares the way" for Darwin (p. 86), whose conception of evolution and the survival of the fittest is in many ways the linchpin for Vaihinger's whole system. That system treats human thought functionally, as an evolving biological phenomenon driven by the struggle for existence. In the introduction, prepared specially for the Ogden translation, Vaihinger writes, "[A]ll thought processes and thought-constructs appear a priori to be not essentially rationalistic, but biological phenomena. . . . Thought is originally only a means in the struggle for existence and to this extent a biological function" (p. xlvi). Here Vaihinger shows his engagement with Nietzsche, and the will to power, just as similar evolutionary commitments expressed by John Dewey show his engagement with Hegel.

6. WOULD IT BE A MIRACLE?

Throughout the whole discussion concerning a shift in status between hypotheses and fictions Vaihinger proceeds, as a constructivist might, without attending to the difference between something *being* a hypothesis or *being* a fiction, and our *believing* it to be so. He also does not face up to the question of what becomes of a false hypothesis; does that make it a fiction? This much is clear. If we knowingly retain a false but useful hypothesis, we have a fiction. The informed use, for example, of Galileo's law of free fall, which postulates a constant gravitational attraction, to calculate free-fall time (or distance) would amount to a fiction, because gravity is not really constant. Similarly with the law of the simple pendulum, the perfect gas law, and so on. In all these cases where local approximations are used, in Vaihinger's terms we have a fiction. In the case of the perfect gas law, we may even have a genuine fiction. In highlighting the idealizations and approximations commonly used in modeling physical phenomena, Vaihinger's central concern is to undo the opinion that if constructs are devoid of reality they are also devoid of utility. Put the other way around, Vaihinger regards the inference from utility to reality as fundamentally incorrect. Thus, despite his pragmatic emphasis on thought as a tool for action, he wants to distinguish his position from the Jamesian form of pragmatism that regards truth to be whatever turns out to be "good" by way of belief, for all the scientific fictions satisfy this formula. On the other hand, what concerns him more is to demonstrate, by the sheer number and range of his examples, that the inference from scientific success at the instrumental level to the literal truth of the governing scientific principles is thoroughly fallacious. I would conjecture that part of the revolutionary impact of Vaihinger's work, and a source of some antagonism to it, was to wean his generation away from what we now call the explanatory argument for realism, or (what Putnam,

I believe, dubbed) the "wouldn't it be a miracle argument." As Vaihinger puts it, "Man's most fallacious conclusion has always been that because a thing is *important* it is also *right*" (p. 69). Vaihinger was concerned that we see through this way of thinking not only in the scientific domain, but more importantly when it comes to religion. For Vaihinger took the idea of God, and the related conceptions of salvation, judgment, immortality, and the like, as fictional (indeed as genuine fictions) and yet of supreme importance in our lives, arguing that this was Kant's way too. In this area, as in the scientific, what Vaihinger preached was critical tolerance, and not skepticism.

Vaihinger, then, offers no general method, no magic criterion, that will enable one to tell whether a construct corresponds to reality or whether a principle is a hypothesis. To the contrary, his examples show that favorites of the realism game (in his time and ours), like fertility or unifying power, are not a reliable guide. Each case has to be looked at on its own, and only time and sensible judgment will tell, if anything will. Vaihinger does, however, suggest a rule of procedure, namely, that we begin by supposing that we are dealing with a fiction and then go from there. Thus Vaihinger would put the burden of proof on the nonfictional foot. Part of what dictates this strategy is Vaihinger's view that we generally find it hard to tolerate the ambiguity and resultant tension that comes from acting on what we acknowledge to be a fiction. There is, he thinks, a natural psychological tendency to discharge this tension (and achieve "equilibrium") by coming to believe that some reality may actually correspond to our useful fictive constructs—"the *as if* becomes *if*"(p. 26). Intellectual integrity, however, requires that we recognize the tendency to believe too readily and suggests a countervailing strategy to help keep us honest. Intellectual growth of the species requires that we come to terms with the tension and learn to tolerate the ambiguity. As with the historic changes that involve a public understanding of the functions and abuses of social and religious dogma, to acknowledge fictive thinking and adopt the strategy of placing the burden on the realist foot would also be a move toward liberation. Although Vaihinger never makes reference to James's will to believe, it is clear that he would have opposed it.

7. CONTRADICTIONS AND GENUINE FICTIONS

There is a sliding line between hypotheses and the semi-fictions. But the genuine fictions are something else, because they are supposed to involve a contradiction and so could not possibly be exemplified in nature. It is this conception of Vaihinger's that has generated the most controversy. M. R. Cohen simply asserts that "in every case the claim of self-contradiction rests on positive misinformation." Indeed, Cohen believes this must be so for otherwise, "no fruitful consequences could be drawn from them [i.e., the so-called genuine fictions] and they would not have the explanatory power which makes them so useful in science" (Cohen, 1923, p. 485).

There are, I think, two related problems here. The first is how the use of a contradiction could be fruitful. The second problem concerns what it means to treat something "as if" it had contradictory properties. Thus the first issue, say for the fiction of atoms, is how they could be useful in treating chemical compounds. The second concerns what it means to treat matter as if it were composed of atoms, when the very concept of an atom is supposed to be contradictory.

These problems are not new, and certainly not peculiar to Vaihinger. For instance, in the realm of mathematical entities, Vaihinger acknowledges the lead of Berkeley, who held that it was impossible for lines and figures to be infinitely divisible. Yet Berkeley also held that the applicability of these geometrical notions required that we speak of lines on paper, for example, as though they contained parts that they really do not; that is, we treat them as though they were divisible. This is precisely Vaihinger's "as if," marking in this case a genuine fiction.

Cohen's way of dealing with these problems is by denial. That is, he just takes it as given that nothing fruitful could be obtained from a contradiction, and that no sense attaches to treating something "as though" it had contradictory properties. (Beineberg: "It's as though one were to say: someone always used to sit here, so let's put a chair ready for him today too, even if he has died in the meantime, we shall go on behaving as if he were coming." Törless: "But how can you when you know with certainty, with complete mathematical certainty, that it's impossible."[13]) Cohen concludes that Vaihinger's examples of genuine fictions, insofar as they are useful, *must* all be mistaken. At best, they are all semi-fictions. This conclusion is drawn too quickly, and apart from the brief examination of one case (that of $\sqrt{-1}$, which is the subject of the parenthetical dialogue), Cohen offers no argument whatsoever for it.

Vaihinger is well aware of the tendency to see difficulties here, which he describes as a "pardonable weakness" (p. 67). Again he refers to the "psychical tension" that may be created by the "as if" (p. 83). (Törless: "[I]t's queer enough. But what is actually so odd is that you can really go through quite ordinary operations with imaginary or other impossible quantities, all the same, and come out at the end with a tangible result! . . . That sort of operation makes me feel a bit giddy.") In mathematics, especially, the use of genuine fictions, according to Vaihinger, requires that we compensate for the contradiction by what he calls (p. 109ff.) the method of "antithetic errors." Vaihinger repeats Berkeley's slogan that "thought proceeds to correct the error which it makes" (p. 61). After drawing what Vaihinger calls "the necessary consequences" the impossible fictional premise just drops out, like the middle term in a syllogism. (Beineberg: "Well, yes, the imaginary factors must cancel each other out in the course of the operation . . .") We use the "as if," Vaihinger also says, to construct a scaffolding around reality which we then cast off when its purpose has been fulfilled (pp. 68–69). (Compare Törless: "Isn't that like a bridge where the piles are there only at the beginning and at the end, with none in the middle, and yet one crosses just as surely and safely as if the whole of it were there?")

With these various caveats, Vaihinger is not trying to explain how contradictions can be fruitful, so much as to reject the idea that some one generic explanation of their fertility is required. (Beineberg: "I'm not going to rack my brains about it: these things never get one anywhere.") He only wants to help free us from the idea that fictional thought is diseased, and that "impossible" concepts are somehow beyond the rational pale. This practice is not much different from what Wittgenstein came to later. Like the later Wittgenstein, Vaihinger thinks if we keep in mind our human purposes we will see that by and large our ordinary ways of thinking, which involve a large amount of fictive activity, are all right. (He does remark that the wish to understand the world as a whole is childish, because when we apply the usual categories outside of their customary home in human experience they engender illusory problems—like that of the purpose of it all [pp. 172–173]. Language on holiday?) Thus Vaihinger tries to slip out of the knot posed by the use of contradictions without suggesting a general answer to the sense of the "as if," when a contradiction is involved, or proposing a theory to answer the Kantian question of how genuine fictions could be useful.

What Vaihinger seems to be suggesting is that one ought to reject these questions about contradictions. Instead of pursuing these "how is it possible that" questions, simply pay detailed attention to reflective practice and notice how fictions are actually used. This optimistic naturalism is tutored by the fruitful history of a long (and, arguably, inconsistent) mathematical practice in the use of infinitesimals. The lesson that Vaihinger takes from this history is that we human animals learn by practice what can be done in a particular domain, and what must be avoided, in order to obtain results that are useful for our purposes. These days one could point to the (inconsistent) use of delta functions in quantum calculations (despite their ex post facto rationalization in terms of Schwartz distributions), or the hierarchy of classifications that allow for the cancellation of infinities (antithetical errors!) in quantum field theory. So what Vaihinger suggests, by example, is that we ought not demand a general theory concerning the fruitful use of contradictions as such. It is sufficient to attend to the variety of successful practices, each of which, literally, shows how it is done.[14]

Although no logic of the use of contradictions comes out of Vaihinger's procedures, there is a conception of rationality. The picture that Vaihinger suggests is that human thinking is circuitous and will try many "roundabout ways and by-paths" (p. xlvii) to find something that works. Rationality, then, simply constrains this activity by imposing the general criterion of fertility. As John Dewey put it, "To suggest that man has any natural propensity for a reasonable inference or that rationality of an inference is a measure of its hold on him [is] grotesquely wrong" (Dewey, 1916, p. 425). For Vaihinger and

Dewey, there is no pre-established harmony between rationality and reality. Rather, rationality is a canon which we assemble bit by bit from experience in order to check our propensity for varied and erroneous inference.

8. ALLIES AND DIFFICULTIES

Once the new positivism took a sharply logical turn (in the hands of Schlick and, more especially, Reichenbach and Carnap) it is not difficult to understand logical positivism's impatience with Vaihinger. His distinctions are many (I count 30 types of fictions, but that is probably not all) and scarcely crisp. Indeed, one can readily picture his relaxed tolerance toward the use of contradictions driving Carnap to distraction, all by itself—as Wittgenstein's similar attitude seems to have done. One wonders, however, about Neurath who always had more historicist leanings. Neurath's work shares several aspects with Vaihinger's, including (1) a community or social conception of language, which Neurath began to champion in the period when Vaihinger's book was first published, and which Neurath later urged on Carnap, (2) Neurath's opposition to "pseudo-rationalism," what we today would describe as a penchant for global meta-narratives about science—his charge against Popper's falsificationism and also against the generalizing thrust of his positivist colleagues, (3) Neurath's naturalism, which made taxonomy part of his philosophical practice, that is, recording, classifying, and (provisionally) accepting the beliefs and behavior of scientists, repairing the boat while at sea, and (4) Neurath's pragmatic tolerance for the use of expedient means and his positive inclination to find some good in each among competing scientific ideas.[15]

Like Neurath, in this positive inclination, I have been trying to emphasize the good things in Vaihinger, in part by way of trying to understand what made his work of such interest in its time. In so doing, however, I need to acknowledge that there are difficulties in Vaihinger's work beyond his fuzziness. Two are conspicuous. One is that Vaihinger frequently relies on secondary sources in place of giving his own, original analysis. This problem stands out, as Cohen suggests, in the rationale he provides for the alleged contradictions in his examples of genuine fictions. For example, a case can be made to sustain Vaihinger's judgment that the number systems constitute genuine fictions, but not by citing Berkeley, or others, without supplementing that with more analysis than Vaihinger himself provides (for he does provide some). A second prominent difficulty is the way in which Vaihinger sets his discussion of fictive judgments in the context of a universal psychologistic logic. Although he is careful to avoid the conception of the mind as a substance, opting for a more behaviorist picture, he does advance the idea that there is a general set of individual, associationist mechanisms involved in the production of fictions. Indeed he sketches some, including a "law of ideational shifts" (p. 124ff.). Clearly the conception of a special psychological faculty for fictive judgment suited his purpose of promoting the value (indeed the necessity) of fictions, and it outweighed his otherwise sensible resistance to proposing perfectly general theories. No doubt, quite apart from Vaihinger's special interest, the idea of universal mechanisms of cognition is also independently appealing. In Vaihinger's time (as in our own), few managed to break free of this sort of psychologism. Vaihinger was not one of them.

There is a third charge against Vaihinger, and the one most frequently encountered. It is that, besotted by the topic of fictions, he simply saw them everywhere (forces, atoms, the classical laws of motion, virtually all of mathematics—not to mention substance, free will, God, and so on) and that, taken to its logical conclusion, his system actually demands this. Concerning this latter charge, consider his distinction between hypotheses and fictions. That depends on contrasting discovery (hypothesis) with invention (fiction), verification (hypothesis) with utilitarian justification (fiction). But why, one might ask, should we take these contrasts as anything more than fictional themselves, as useful expedients in the labor of the mind?[16] What gives the question its bite is that Vaihinger provides no firm grounds for sorting and grading into fictional versus nonfictional. It is part of his own scheme that there are no general answers to such questions: that is, answers that can be derived from general subject-neutral principles. His answer about how to grade and sort, if there is one, has to be narrowly tailored, topic specific, and historicist. That is what his system demands. To charge, on that account, that Vaihinger has no answer and that therefore we are free to assimilate all thought to fiction is to impose on Vaihinger the nihilistic standard according to which if a question cannot be answered on the basis of perfectly general principles, then we are free to answer it just how we please. Because Vaihinger's whole enterprise stands in opposition to this nihilism, however, this is not an appropriate standard for the interpretation of his work. Thus, instead of showing up a deep flaw in Vaihinger, this objection exposes the tenacity with which one can cling to misplaced presuppositions about generality. This nicely illustrates what Vaihinger calls "the preponderance of the means over the ends" (which we know as Parkinson's Law). In this case, that preponderance accounts for how generality, a means to certain ends, gets displaced in the course of time into an end-in-itself. The remaining charge, that Vaihinger simply finds altogether too many fictions, has, of course, to be carefully defended and responded to on a case by case basis. Vaihinger would ask no more, nor should we damn him on the basis of anything less.

9. VAIHINGER'S REPUTATION

Despite its moderately historical veneer, my treatment of Vaihinger has been rather unhistorical. I have not asked who his readers were, with what themes of theirs his work resonated, into what larger cultural phenomena, trends, and institutions it fit, and so on. Moreover, although I believe that Vaihinger's ideas on fictions were widely influential and I can readily point to similar ideas similarly expressed over a broad intellectual terrain, I have not really traced the vectors by virtue of which those ideas were spread there. Similar social-historical work needs to be pursued to understand why Vaihinger fell out of favor just when he did, and I have not done that work either. Nevertheless, I want to address this last issue, as it relates to logical positivism, where some things can be said for sure and some conjectures can at least be raised.

For sure, Vaihinger's flaws alone do not explain the disrepute into which he has fallen. After all, of what is he guilty? He makes a lot of distinctions, difficult to keep in mind, which are not even exhaustive and complete. He is somewhat fuzzy, sometimes. He sometimes comes up short on analysis. Naively, he accepts a lot of what scholars have said, and scientists done, as correct. He subscribes to an individualistic picture of cognition that is more universal that he ought to allow. He may have made some factual errors in what he took to be fictions, and what not. I might add that he has a somewhat old-fashioned penchant for coining "laws" and for diagnosing historical trends. He also likes to moralize. Sometimes he repeats himself. In a large work, no doubt there are some inconsistencies. (I have not found a good example of this, unlike Schlick, apparently; but let us grant it.)

Examining this list, it looks like what Vaihinger mostly needed was a competent editor! Yet what he received from the logical positivists, and his legacy from them to us, has been something rather different. Put most simply, the positivists set about making Vaihinger a marginal figure. They succeeded. In today's literature, when he is mentioned at all, it is almost always in the margins—in a footnote one-liner or a parenthetical thrust. One will protest, perhaps, that in compiling my list of flaws I have not discussed the originality, quality, or viability of his ideas. Surely, in the end, that is what counts. Indeed it does, and that is my point. My omission tracks Vaihinger's positivist critics, for they do not discuss the originality, quality, or viability of his ideas either. Mainly they mock him and promptly place his "As If" in the trash. This is especially striking behavior in figures like Carnap, Reichenbach, Schlick, and allies, who were formidable critics. Precisely by not bringing Vaihinger into their critical discussions, the logical positivists made it appear that his ideas were simply not worth discussing. Because, as we have seen, many of his ideas were also theirs, that explanation of why Vaihinger was not treated seriously cannot be correct. So why did they do it? One clue to the dynamic may be found by reflecting on Neurath. Neurath was the organizer of the logical positivist movement, the "big engine" whose energy and political skills kept the group growing and alive. Until a very recent revival of interest in Neurath, however, few have promoted him as a serious thinker on par with the others. Neurath's influence on the intellectual development of logical positivism has lain in the shadows. His ideas, especially those in opposition to the mainstream, were mostly lost from general view. I would suggest two things that might help explain this neglect. First, although Neurath wrote some slogans about verifiability and meaning, he never really made the commitment to an analysis of scientific language that became a characteristic feature of logical positivism.[17] Second, Neurath mostly stood in opposition to those tendencies in the movement that set the projects for understanding science in general terms: the projects of a general account of explanation, a general theory of confirmation, a general account of laws, or theories, and so on. Indeed, what has come down as the modern agenda of philosophy of science includes, for the most part, the very things that Neurath thought not sensible to pursue. As indicated earlier in this chapter, the approach that Neurath advocated was more historicist, taxonomic, and naturalistic than the items on that agenda. His methodology, then, did not give pride of place to formal methods, meaning, and language. His orientation was piecemeal and particularist, self-consciously not global (the approach he mocked as "pseudo-rational"). Obviously, Neurath also had a great deal in common with his positivist colleagues. But, I suggest, by rejecting a linguistic and global orientation Neurath was easily cast as an outsider, and his reservations about these features of the program were by and large just not discussed. This was certainly not due to spite or bad feelings, but because the others had their own projects to pursue and could not be always engaged in justifying their whole approach. Like Einstein with respect to the quantum theory, Neurath suffered the fate of one who keeps questioning fundamentals. After a while, his reservations were set aside in order to get on with things.

My point about Vaihinger should be plain; it is the same point. In contrast with Scheffler's postwar setting for fictionalism (see section 1), Vaihinger was more like Neurath in not having a linguistic and meaning-related orientation, despite some occasional philological excursions. Unlike Scheffler, he was also by and large a confirmed taxonomist and naturalist of science, not subscribing to the global projects so dear to the neo-positivist hearts. Like Neurath, these aspects of Vaihinger could not be confronted without bringing into question the whole logical positivist project. So, Vaihinger was not discussed. He was written off. In Vaihinger's case, however, there was a further problem. For, as we have seen, Vaihinger had actually anticipated and set up institutional structures to pursue a program of philosophical reform and re-evaluation uncomfortably close to the project of logical positivism. Logical positivism, however, proclaimed itself a new program in the history of thought, the vanguard of a new enlightenment, the cutting edge of a new modernity. To be sure, the movement claimed the heritage of Mach and Poincaré, of Helmholtz, Hilbert, Einstein, Bohr, and (sometimes) Freud. These, however, were leading scientists of the era. With regard to existing philosophical schools, however, logical positivism acknowledged no peers, presenting itself as a fresh starting point. (Indeed, even their debt to Kant and their neo-Kantian contemporaries has to be excavated.) To cultivate the image of new-philosophical-man, they had to downplay the continuities between them and the popular Vaihinger, distancing themselves from him. So they did.

10. VAIHINGER'S LEGACY

Perhaps the dismissive attitude that logical positivism adopted to Vaihinger was overdetermined by the social and political circumstances suggested earlier, and probably by others as well. Whatever a better historical treatment would show, our attitude need not be bound by the judgment of those other times. With respect to general philosophical orientation, Vaihinger (and Neurath) points us toward a more naturalistic and particularistic approach to understanding science, an approach (like Vaihinger's to religion) that is at once critical and tolerant (with strong emphasis on both). This approach calls into question the viability of the universal projects of philosophy of science, demanding a hearing for a non-theory-dominated way. It also moves us away from a preoccupation with language and meaning. The approach is, broadly speaking, pragmatic in its emphasis on the importance of scientific practice in relation to scientific theory, although it is not reductive in the Jamesian way with respect to truth. This general sort of orientation is already being explored in various contemporary naturalisms, in my NOA, and also in some constructivist and broadly deflationist programs. I take these to be part of Vaihinger's legacy. What then of fictions and the "As If"?

Vaihinger's emphasis on fictions exalts the role of play and imagination in human affairs. He finds no realm of human activities, even the most serious of them, into which play and imagination fail to enter. Surely he is right. These faculties are part of the way we think ("constructively"), approach social and intellectual problems ("imaginatively"), employ metaphor and analogy in our language, and relate to others every single day.

Within science, idealizations and approximations are an integral part of ordinary everyday procedure. The representation of three dimensions on two (i.e., graphing), the conceptualization of four (or 27!) in terms of three, all call on the imagination to create a useful fiction—as does any pictorial presentation of data. The images by virtue of which whole fields are characterized ("black hole," "strings," "plates," "bonds," "genetic code," "software," "systems," "chaos," "computable," "biological clock," and so on) have the same character. Indeed, new techniques are constantly being developed for the creation of scientific fictions. Game and decision theory come readily to mind. Computer simulation, in particle physics or weather forecasting, is also a significant postwar example. Preeminently, the industry devoted to modeling natural phenomena, in every area of science, involves fictions in Vaihinger's sense. If you want to see what treating something "as if" it were something else amounts to, just look at most of what any scientist does in any hour of any working day.

In these terms, Vaihinger's fictionalism and his "As If" are an effort to make us aware of the central role of model building, simulation, and related constructive techniques, in our various scientific practices and activities. Vaihinger's particularist attitude over the question of whether and to what extent any model captures an element of the truth warns us to be wary of overriding arguments about how to interpret (useful) scientific constructs in general. History shows us that there are no magical criteria that fix the interpretation, and no simple answers here. (History also shows that in the puzzle cases that exercise philosophers, where ordinary scientific procedures do not seem to settle the issue, science gets along perfectly well when these realist questions are not pursued. But emphasizing this fact would be to push Vaihinger toward NOA.)

By distancing itself from Vaihinger, logical positivism missed an opportunity that would have kept it in the mainstream of scientific thought throughout this century. Those who would dismiss a view by associating it with Vaihinger and his "As If" make the same error. For the dominant self-conception of postwar science has been that of science as the builder of useful models. In our century Vaihinger was surely the earliest and most enthusiastic proponent of this conception, the preeminent 20th-century philosopher of modeling.

NOTES

1. An earlier version of this chapter was presented to a conference at Princeton University organized by Bas van Fraassen in the spring of 1992, where there were many useful comments. Joseph Pearson and Thomas Uebel helped with the history of Vaihinger and his fictions. Mara Beller provided me with a copy of the Schrödinger letter from which the epigraph is drawn and with encouragement to pursue my interest in Vaihinger. Thomas Ryckman offered good criticism and advice. Thanks, all!
2. Thus Horwich (1991) conjures up Vaihinger's fictionalism in order to set the tone for a criticism of van Fraassen's constructive empiricism.
3. For fictionalism in mathematics see Papineau (1988), whose position resembles that of Berkeley and Vaihinger, although Papineau is not concerned to trace these antecedents.
4. Page number references to Vaihinger are to the Ogden translation (Vaihinger, 1924).
5. Ryckman (1991) notes some reversals in Schlick's attitude, which was receptive to Vaihinger early on but then turned hostile.
6. For details about the history of *Erkenntnis*, see Hempel (1975), and for its relation with Vaihinger and the *Annalen*, see Hegselmann and Siegwart (1991).
7. See Gould (1970).
8. Spariosu (1989) discusses Vaihinger and draws out parallels between his fictionalism and the ideas expressed by leading thinkers in German-speaking physics prior to the Second World War. Unfortunately, many of Spariosu's

philosophical or scientific conclusions and generalizations seem unsound. Still, the textual record of Vaihingerisms that he compiles is impressive evidence of the resonance of fictionalism among prominent scientists of the time. One could add the complementarity of Niels Bohr to Spariosu's list. See the epigraph.

9. The legal realism of Jerome Frank (1970), for example, makes extensive use of Vaihinger's ideas.

10. "That which is not useful is vicious." Attributed to Cotton Mather.

11. For discussions of Vaihinger and the Kantian "thing-in-itself," especially with respect to the third *Critique,* see Schaper (1965, 1966).

12. Also important are what he calls dogmas, especially given his religious and ethical concerns. I will omit his treatment of dogmas, given our concerns.

13. This citation and the others running parenthetically through this section are from Musil (1906, pp. 106–107) whose dialogue between Beineberg and Törless struggles with exactly the issues that divide Cohen from Vaihinger, and even over Cohen's one case of imaginary numbers. Musil provides illustrations for arguments that Cohen did not find. (My thanks to Joseph Pearson for calling my attention to these passages in Musil.)

14. David Lewis (1982; 1983, pp. 276–278) pursues a contrary tack, suggesting a quite general approach via "disambiguation" for how contradictions may fruitfully be used without harm.

15. See Uebel (1992) for a discussion of these and other features of Neurath's work.

16. This is a line pursued by Spariosu (1989).

17. Appearances to the contrary, his proposals about protocol sentences were not in aid of regimenting the language of science. Rather, Neurath was trying to display the complex range of objective, subjective, and social factors that enter into any scientific report, in part to undercut the idea that they *could* usefully be regimented. See Uebel (1991).

3 Laboratory Fictions

Joseph Rouse

1. INTRODUCTION

When philosophers consider fictions or fictional representations in science, we typically have in mind some canonical cases: idealized models, simulations, thought experiments, or counterfactual reasoning. The issue raised by scientific fictions may also seem straightforward. Sciences aim to discover actual structures and behaviors in the world, and to represent and understand them accurately. A role for scientific fictions provokes the question of how fictional representations, or, more provocatively, misrepresentations could contribute to scientific understanding of how the world actually is.

I will address a different issue, concerning conceptual meaning and significance rather than truth or falsity. The two issues are closely connected, but it is a mistake to conflate them. Two decades ago, Nancy Cartwright collected some important essays under the provocative title *How the Laws of Physics Lie* (1983). Laws lie, she argued, because they do not accurately describe real situations in the world. Descriptions of actual behavior in real situations require supplementing the laws with more concrete models, ad hoc approximations, and ceteris paribus provisos. Indeed, Cartwright argued that the fictional character of physical laws was analogous to literary, or more specifically, theatrical fiction; like film or theatrical productions, we might say, the genre of physical law demands its own fictive staging.

Shortly thereafter, I argued (1987, chap. 5) that Cartwright had mischaracterized the import of her concerns. Her arguments challenge the *truth* of law-statements only if their *meaning* were fixed in ways at odds with the actual use of such expressions in scientific practice. After all, the need for models, provisos, and ad hoc approximations to describe the actual behavior of physical systems in theoretical terms comes as no surprise to physicists. The models were integral to their education in physics, and the open-ended provisos and approximations needed to apply them were implicit in their practical grasp of the models. Thus, the "literal" interpretation of the laws that Cartwright once took to be false does not accurately express what the laws mean in scientific practice. [1]

I would now express this point more generally. The meaning of an expression such as $F = ma$ or one of Maxwell's equations is a normative matter, expressing a connection between the ways and circumstances in which that expression is appropriately employed in scientific practice, and the consequences that appropriately follow from its employment.[2] To that extent, of course, understanding laws and similar verbal or mathematical expressions cannot be easily disentangled from understanding the circumstances to which they apply. As Donald Davidson expressed the more general point, we thereby "erase the boundary between knowing a language and knowing our way around in the world generally" (1986, pp. 445–446).

Cartwright (1999a) herself now makes a related point about the tradeoffs between truth and meaning. What concerns her is not the truth of laws, but their scope: Which events should we actually take them to be informative about and accountable to? Moreover, she equates the scope of the laws with the scope of the concepts they employ. The laws of classical mechanics apply wherever the causal capacities that affect motions are appropriately characterized as "forces." I am not here concerned with Cartwright's answer to the question of which circumstances fall within the domain of the concept of "force." I only want to insist that while questions of meaning and of truth are interconnected, they must remain distinct. We cannot ask whether a theory, a law, or any other hypothesis is true unless we have some understanding of what it says, and to which circumstances it appropriately applies.

To discuss conceptualization and meaning, I will consider a very different kind of case than the canonical examples of "scientific fictions." I have in mind the development and exploration of what I once called laboratory "microworlds" (Rouse, 1987), which Hans-Jörg Rheinberger (1997) has since characterized as "experimental systems." I once described "microworlds" as "systems of objects constructed under known circumstances and isolated from other influences so that they can be manipulated and kept track of, . . . [allowing scientists to] circumvent the complexity [with which the world more typically confronts us] by constructing artificially simplified 'worlds'" (1987, p. 101). Some illustrative experimental microworlds include the Morgan group's system for mapping genetic mutations in *Drosophila melanogaster*, the many setups in particle physics that direct a source of radiation toward a shielded target and detector, or the work with alcohols and their derivatives that Ursula Klein (2003) has shown to mark the beginnings of experimental organic chemistry. These are not verbal, mathematical, or pictorial representations of some actual or possible situation in the world. They are not even physical models, like the machine-shop assemblies that Watson and Crick manipulated to represent three-dimensional structures for DNA. They are instead novel, reproducible arrangements of some aspect of the world.

My consideration of experimental systems in relation to the canonical scientific fictions may seem strange. Discussions of scientific fictions commonly take experimentation for granted as relatively well understood. Against this background, the question is sometimes raised of whether thought experiments or computer simulations relevantly resemble experimental manipulations as "data-gathering" practices.[3] I proceed in the opposite direction, asking whether and how the development of experimental systems resembles the construction and use of more canonical kinds of "scientific fiction." The issue is not whether simulations or thought experiments can be sources of data, but whether and how laboratory work takes on a role akin to that of thought experiments in articulating and consolidating conceptual understanding. Philosophers have tended to exclude experimentation from processes of conceptual development. To caricature quickly a complex tradition, the logical empiricists confined experimentation to the context of justification rather than discovery; postempiricists and scientific realists emphasized that experimentation *presupposes* prior theoretical articulation of concepts, whereas reaction to the excesses of both traditions proclaimed that experimentation has a life of its own apart from developing or testing concepts and theories. None of these traditions has said enough about experimentation as itself integral to conceptual articulation.

As background to what I will say about experimental systems, consider briefly Kuhn's classic account of the function of thought experiments. Thought experiments become important when scientists "have acquired a variety of experience which could not be assimilated by their traditional mode of dealing with the world" (Kuhn, 1977, p. 264). By extending scientific concepts beyond their familiar uses, he argued, thought experiments bring about a conceptual conflict rooted in those traditional uses, rather than finding one already implicit in them. Kuhn insisted upon that distinction, because he took the meaning of concepts to be open-textured rather than fully determinate. By working out how to apply these concepts in new, unforeseen circumstances, thought experiments retrospectively transformed their use in more familiar contexts, rendering them problematic in illuminating ways.

Thought experiments could only play this role, however, if their extension to the newly imagined setting genuinely extended the original, familiar concepts. Kuhn consequently identified two constraints upon such imaginative extension of scientific concepts, "if it is to disclose a misfit between traditional conceptual apparatus and nature": First, "the imagined situation must allow the scientist to employ his usual concepts in the way he has employed them before, [not] straining normal usage" (1977, pp. 264–265). Second, "though the imagined situation need not be even potentially realizable in nature, the conflict deduced from it must be one that nature itself could present; indeed, . . . it must be one that, however unclearly seen, has confronted him before" (1977, p. 265). Thought experiments, that is, are jointly parasitic upon the prior employment of concepts, and the world's already-disclosed possibilities; like Davidson, Kuhn found it hard to disentangle our grasp of concepts from "knowing our way around in the world." Against that background, thought experiments articulate concepts by presenting concrete situations that display differences that are intelligibly connected to prior understanding. In Kuhn's primary example, the difference between instantaneous and average velocity only becomes conceptually salient in circumstances where comparisons of velocities in those terms diverge. My question, however, is how scientific concepts come to have a "normal usage" in the first place, not merely acquainting us with what actually happens, but providing a grasp of possibilities, and thus the situations "that nature itself could present."

The novel circumstances presented in experimental systems or thought experiments are important because they make salient a conceptually significant difference that does not show itself clearly in more "ordinary" circumstances. Yet experimental systems sometimes play a pivotal role in making possible the conceptual articulation of a domain of phenomena in the first place. A postempiricist commonplace rejects any "Whig" history of science that narrates a relatively seamless transition from error to truth. Yet in many scientific domains, earlier generations of scientists could not yet have erred, because the possibility of error was not yet open to them. In the most striking cases, scientists' predecessors either had no basis whatsoever for making claims within a domain, or could only make vague, unarticulated claims. In Hacking's (1984) apt distinction, they lacked not truths, but possibilities for truth or error: They had no way to reason about such claims, and thus could not articulate claims that were "true-or-false." A distinctive feat of laboratory science, then, is to allow new aspects of the world to show up as conceptually articulable.

2. PHENOMENA AND CONCEPTUAL ARTICULATION

To understand how the construction of experimental systems plays such a role in conceptual articulation, consider this remark by Hacking about how scientists come to "know their way around in the world generally":

> In nature there is just complexity, which we are remarkably able to analyze. We do so by distinguishing, in the mind, numerous different laws. We also do so by presenting, in the laboratory, pure, isolated phenomena. (1983, p. 226)

Hacking's conception of phenomena is now familiar. He was talking about events in the world rather than appearances to the mind, and he argued that most phenomena were created in the laboratory rather than found in the world. Experimental work does not simply strip away confounding complexities to reveal underlying nomic simplicity; it creates new complex arrangements as indispensable background to any foregrounded simplicity. Yet most philosophical readers have not taken Hacking's suggested parallel between phenomena and laws as modes of analysis sufficiently seriously. We tend to think only laws or theories allow us to analyze and understand nature's complex occurrences. Creating phenomena may be an indispensable means to discerning relevant laws or constructing illuminating theories, but they can only indicate possible directions for analysis, which must be developed theoretically. I will argue, however, that it is a mistake to treat laboratory phenomena in this way as merely indicative *means* to the verbal or mathematical articulation of theory, even if one acknowledges that experimentation also has its own ends. Experimental practice can be integral rather than merely instrumental to theoretical understanding.

As created artifacts, laboratory phenomena and experimental systems have a distinctive aim. Most artifacts, including the apparatus within an experimental system, are used to accomplish some end. The end of an experimental system, however, is not so much what it does, as what it shows. Experimental systems are novel rearrangements of the world that allow some aspects that are not ordinarily manifest and intelligible to *show* themselves clearly and evidently.

Sometimes such arrangements isolate and shield relevant interactions or features from confounding influences. Sometimes they introduce signs or markers into the experimental field, such as radioactive isotopes, genes for antibiotic resistance, or correlated detectors for signals whose conjunction indicates events that neither signifies alone. This aspect of experimentation reverses the emphasis from traditional empiricism: What matters is not what the experimenter observes, but what the phenomenon shows.

Catherine Elgin (1991, this volume) usefully distinguishes the features or properties an experiment *exemplifies* from those that it merely *instantiates*. In her example, rotating a flashlight 90 degrees merely instantiates the constant velocity of light in different inertial reference frames, whereas the Michelson/Morley experiment exemplifies it.[4] Elgin thereby emphasizes the symbolic function of experimental performances, and suggests parallels between their cognitive significance and that of paintings, novels, and other artworks. She notes that a fictional character such as Nora in *A Doll's House* strikingly exemplifies a debilitating situation that the lives of many actual women in conventional bourgeois marriages merely instantiate. Yet Elgin still distinguishes scientific experimentation from both literary and scientific fictions. An experiment actually instantiates the features it exemplifies, whereas thought experiments and computer simulations share with many artworks the exemplification of features they instantiate only metaphorically.

Elgin's distinction between actual experiments and fictional constructions gives priority to instantiation over exemplification. Nora's life is fictional, and is therefore only metaphorically constrained, whereas light within the Michelson interferometer really does travel at constant velocities in orthogonal directions. Thought experiments, computer simulations, and novels are derivative, fictional, or metaphorical exemplifications, because exemplifying a conceptually articulated feature depends upon already instantiating that feature. The feature is already 'there' in the world, awaiting only the articulation of concepts that allow us to recognize it. Unexemplified and therefore unconceptualized features of the world would then be like the statue of Hermes that Aristotle thought exists potentially within a block of wood, whose emergence awaits only the sculptor's (or scientist's) trimming away of extraneous surroundings.[5]

In retrospect, with a concept clearly in our grasp (or better, with ourselves already in the grip of that concept), the presumption that the concept applies to already-extant features of the world is unassailable. Of course there were mitochondria, spiral galaxies, polypeptide chains, and tectonic plates before anyone discerned them, or even conceived their possibility. Yet this retrospective standpoint, where the concepts are already articulated and the only question is where they apply, crucially mislocates important aspects of scientific research. In Kantian terms, researchers initially seek reflective rather than determinative judgments. Scientific research must articulate concepts with which the world can be perspicuously described and understood, rather than simply apply those already available. To be sure, conceptual articulation does not begin de novo, but extends a prior understanding that gives indispensable guidance to inquiry. Yet in science, one typically recognizes such prior articulation as tentative and open-textured, at least in those respects that the research aims to explore.

The dissociation of experimental work from conceptual articulation reflects a tendency to think of conceptual development as primarily verbal, a matter of gaining inferential control over the relations among our words. Quine (1953, p. 42) encapsulated that tendency with his images of conceptual schemes as self-enclosed fabrics or fields that accommodate the impact of unconceptualized stimuli at their boundaries solely by internal adjustments in the theory. Both Donald Davidson (1984) and John McDowell (1994) have criticized the Quinean image, arguing that the conceptual domain is unbounded by anything "extra-conceptual." I agree, yet reflection upon the history of scientific experimentation strongly suggests the inadequacy of Davidson's and McDowell's own distinctive ways of securing the unboundedness of the conceptual.[6] Against Davidson, that history reminds us that conceptual articulation is not merely intralinguistic.[7] Against McDowell, the history of experimentation reminds us that conceptual articulation incorporates causal interaction with the world, and not just perceptual receptivity.

Both points are highlighted by examples in which experimentation opened whole new domains of conceptual articulation, where previously there was, in Hacking's apt phrase, "just complexity." Think of genes before the Morgan group's correlations of crossover frequencies with variations in chromosomal cytology (Kohler, 1994); of heat and temperature before the development of intercalibrated practices of thermometry (Chang, 2004); of interstellar distances before Leavitt's and Shapley's tracking of period–luminosity relations in Cepheid variables; of the functional significance of cellular structure before the deployment of the ultracentrifuge and the electron microscope (Bechtel, 1993; Rheinberger, 1995); or of subatomic structure before Rutherford targeted gold leaf with beams of alpha particles. These features of the world were less ineffable than the "absolute, unthinkable, and undecipherable nothingness" that Hacking (1986) memorably ascribed to anachronistic human kinds. They nevertheless lacked the articulable differences needed to sustain conceptual development. What changed the situation was not just new kinds of data, or newly imagined ways of thinking about things, but new interactions that articulate the world itself differently. For example, surely almost anyone in biology prior to 1930 would have acknowledged that cellular functioning requires a fairly complex internal organization of cells in order for them to perform their many roles in the life of an organism. Yet such acknowledgment was inevitably vague and detached from any consequent program of research (apart from the identification of some static structures such as nuclei, cell walls, mitochondria, and a few recognized in vitro biochemical pathways). Without further *material* articulation of cellular components, there was little one could say or do about the integration of cellular structure and function.

The construction of experimental microworlds thus plays a distinctive and integral role in the sciences. Heidegger, who was among the first to give philosophical priority to the activity of scientific research over the retrospective assessment of scientific knowledge, forcefully characterized the role I am attributing to some experimental systems:

> The essence of research consists in the fact that knowing establishes itself as a "forging-ahead" (*Vorgehen*) within some realm of entities in nature or history. . . . Forging-ahead, here, does not just mean procedure, how things are done. Every forging-ahead already requires a circumscribed domain in which it moves. And it is precisely the opening up of such a domain that is the fundamental process in research. (Heidegger, 1950, p. 71; 2002, p. 59, translation modified)

The creation of laboratory microworlds is often indispensable to opening domains in which scientific research can proceed to articulate and understand circumscribed aspects of the world.

3. LABORATORY FICTIONS

What does it mean to open up a scientific domain, and how are such events related to the construction of experimental systems? Consider first that experimental systems always have a broader "representational" import. It is no accident that biologists speak of the key components of their experimental systems as model organisms, and that scientists more generally speak of experimental models. The cross-breeding of mutant strains of *Drosophila* with stock breeding populations, for example, was not a phenomenon of interest for its own sake, but it was also not merely a peculiarity of this species of *Drosophila*. The *Drosophila* system was instead understood, rightly, to show something of fundamental importance about *genetics* more generally; indeed, as I shall argue, it was integral to the constitution of genetics as a research field. Elgin (this volume) reminds us that experiments often play a symbolic role, by exemplifying rather than merely instantiating their features. Domain-opening experimental systems "exemplify" in an even stronger sense, because they help constitute the conceptual space within which their features are intelligible as such.

I suggest that we think of such domain-constitutive experimental systems as "laboratory fictions." It is no objection to this claim that laboratory systems seem to present rather than represent a phenomenon. Their symbolic role in exemplification indicates that some laboratory phenomena also have an intentional directedness beyond themselves. Yet talking about actual circumstances in the laboratory as "fictions" may still seem strange, even when their creation and refinement are unprecedented. By definition an experimental setup is an actual situation, in contrast to thought experiments or some theoretical models, which represent idealized or even impossible situations. Ideal gases, two-body universes, "silogen atoms" (Winsberg, this volume), or observers traveling alongside electromagnetic waves nicely exemplify this apparent contrast between fictional representations and fact. Even among this class of theoretical representations, Winsberg (this volume) argues persuasively for distinguishing some idealized constructions (such as ideal gases or frictionless planes) that serve as reliable guides to an actual domain, and hence should not be regarded as fictional representations, and some (such as silogen atoms) whose functional role in a larger representational system requires that they forego representational reliability within a more localized domain.

Winsberg's distinction between fictional and nonfictional elements of theoretical models turns on how these theoretical constructions fulfill a representational function. Yet representation is not the sole or, I would argue, even the primary function of scientific work. Scientific work also has a discursive function that articulates conceptual norms. Indeed, I have argued elsewhere (Rouse, 2002, chaps. 5–9) that scientific models and theories only have a representational function through their role in discursive practices. My criticism of Cartwright's early work that treated laws as fictions also emphasized situating the representational function of laws within the context of their discursive uses within scientific practice. Moreover, serious scholarly discussion of literature nowadays likewise give priority to the discursive role of literary work, including literary fictions.[8] Within certain contexts, it is undoubtedly appropriate to take for granted the discursive context that allows the representational relation between a theoretical model or concept and some situations in the world to seem transparent. Part of the point of my chapter, however, is to highlight some of the aspects of scientific practice that condition such treatments of scientific representation.

Two interconnected features of the experimental systems that bring an entire field of phenomena "into the open" or "into the space of reasons" lead me to conceive them as laboratory fictions. First is the *systematic* character of experimental operations. Fictions in the sense that interest me are not just any imaginative construction, but have sufficient self-enclosure and internal complexity to constitute a situation whose relevant features can be identified through their mutual interrelations. Second, fictions *constitute* their own "world." The point is not that nothing outside the fictional construction resembles or otherwise corresponds to the situation it constructs, for many fictional constructions do have "real" settings. Even in those cases, however, they constitute a world internally rather than by external references. In this sense, we can speak of Dickens's London in the same way that we speak of Tolkien's Middle Earth.

My first point, that scientific experimentation typically requires a more extensive experimental system rather than just individual experiments, is now widely recognized in the literature.[9] Ludwik Fleck (1979) was among the first to highlight the priority of experimental systems to experiments, and most later discussions follow his concern for the justificatory importance of experimental systematicity: "To establish *proof*, an entire system of experiments and controls is needed, set up according to an assumption or style and performed by an expert" (Fleck, 1979, p. 96, my emphasis). Hacking's (1992) discussion of the "self-vindication" of the laboratory sciences also primarily concerned epistemic justification. The self-vindicating stability of the laboratory sciences, he argued, is achieved in part by the mutually self-referential adjustment of theories and data.

I am making a different claim: It is primarily by creating systematically intraconnected "microworlds" that new domains are opened to contentful conceptual articulation at all.[10] "Genes," for example, were transformed from merely hypothetical posits to the locus of a whole field of inquiry ("genetics") by the correlation of crossover frequencies of mutant traits with visible transformations in chromosomal cytology, in flies cross-bred to a standardized breeding population. Carbon chemistry (as distinct from the phenomenological description of organically derived materials) likewise became a domain of inquiry through the systematic, conceptually articulated tracking of ethers and other derivatives of alcohol (Klein, 2003). Leyden jars and voltaic cells played similar roles for electricity. What is needed to open a novel research domain is typically the ability to create and display an intraconnected field of reliable differential effects: not merely creating phenomena, but creating an experimental practice.

My shift in concern from justification to conceptual articulation and domain-constitution correlates with a second difference from Hacking's account. Hacking sought to understand the eventual stabilization of fields of laboratory work, as resulting from the mutually self-vindicating adjustment of theories, apparatus, and skills. I consider the beginning of this process rather than its end, the *opening* of new domains for conceptual articulation rather than their eventual practical and conceptual stabilization. Yet my argument puts Hacking's account of self-vindication in a new light. We need to consider not only the self-vindicating justification of experimental work, but its scientific significance more generally. However self-referential and self-vindicating a complex of experimental phenomena and theory may be, its place within an ongoing scientific enterprise depends upon its being informative beyond the realm of its self-vindication. Hacking's emphasis upon self-vindicating stabilization thus may subtly downplay the intentional and conceptual character of experimental systems.[11]

To understand the intentionality of experimental systems, I turn to the second feature of "laboratory fictions," their constitutive character.[12] The constitution of a scientific domain accounts for the conceptual character of the distinctions functioning within the associated field of scientific work. Consider what it means to say that the *Drosophila* system developed initially in Morgan's laboratory at Columbia was about *genetics*. We need to be careful here, for we cannot presume the identity and integrity of genetics as a domain. The word 'gene' predates Morgan's work by several years, and the notion of a particulate, germ-line "unit" of heredity emerged earlier in the work of Mendel, Darwin, Weismann, Bateson, and others. The conception of genes as the principal objects of study within the domain of genetics marks something distinctive, however. Prior conceptions of heredity did not and could not distinguish genes from the larger processes of organismic development within which they functioned. What the *Drosophila* system initially displayed, then, was a field of distinctively genetic phenomena, for which the differential development of organisms was part of an apparatus that articulated genes as relative chromosomal locations and characteristic patterns of meiotic crossover, as well as phenotypic outcomes. This example also highlights the significance of my shift from the representational to the discursive function of an experimental system. Morgan's *Drosophila* system does not denote the domain of genetics, any more than Austen's *Pride and Prejudice* denotes the traits of character that it articulates within the fictional context of 'three or four families in a country village' (Elgin, this volume). Austen rightly titled her novel *Pride and Prejudice* rather than *Elizabeth Bennet and Mr. Darcy*; its primary function as a fictional construction is not denotative, and to read it as primarily about these characters would be to miss the point. One would similarly miss the point if one regarded the Morgan group's research as a study of artificially standardized fruit flies. The discursive role of these fictional constructions is articulative rather than denotative; they are "about" an entire interconnected domain of entities, properties, and relations, rather than about any particular entities or properties within that domain.

What the *Drosophila* system did specifically was to allow a much more extensive inferential articulation of the concept of a gene. *Concepts* are marked by their possible utilization in contentful judgments, which acquire their content inferentially.[13] For example, a central achievement of *Drosophila* genetics was the identification of phenotypic traits (or trait-differences) with chromosomally located "genes." Such judgments cannot simply be correlations between an attributed trait and what happens at a chromosomal location, because of their inferential interconnectedness. Consider the judgment in classical *Drosophila* genetics that the Sepia gene is *not* on chromosome 4.[14] This judgment does not simply withhold assent to a specific claim; it has the more specific content that either the Sepia gene has some other chromosomal locus, or that no single locus can be assigned to the distinctive traits of Sepia mutants. Such judgments, that is, indicate a more-or-less definite space of alternatives. Yet part of the content of the "simpler" claim that Sepia is on chromosome 3 is the consequence that it is not on chromosome 4. Thus, any single judgment in this domain presupposes the intelligibility of an entire conceptual space of interconnected traits, loci, and genes (including the boundaries that mark out what is not a relevant constituent of that space).[15]

My point about the relation between scientific domains and entities disclosed within those domains can be usefully highlighted by a brief comparison to Hanna and Harrison's (2002) discussion of the conceptual space of naming. Proper names are often regarded as the prototypical case of linguistic expressions whose semantic significance is directly conferred by their relationship to entities bearing those names. Hanna and Harrison's counterclaim, which I endorse, is that proper names themselves refer not to persons, ships, cities, and the like, but to "nomothetic" objects ("name-bearerships") disclosed within a larger framework of practices that they identify collectively as the "Name-Tracking Network"[16]:

> To give a name is . . . to reveal, in the ordinary way of things, a label that has been used for many years, through occurrences of tokens of it in the context of many naming practices, to trace, or track, one's progress through life. Such tracking operates by way of a variety of practices: the keeping of baptismal rolls, school registers, registers of electors; the editing and publishing of works of reference of the *Who's Who* type, the inscribing of names, with attached addresses, in legal documents, certificates of birth, marriage, and death, and so on. Such practices are mutually referring in ways that turn them into a network through which the bearer of a given name may be tracked down by any of dozens of routes. (Hanna & Harrison, 2002, p. 108)

We use names to refer to persons, cities or ships, but we can do so only via the maintenance and use of such an interconnected field of practices. Such practices constitute trackable "name-bearerships" through which names can be used *accountably* to identify human beings or other entities.

Experimental systems play a role within scientific domains comparable to the role of the Name-Tracking Network in constituting name-bearerships. They mediate the accountability of verbally articulated concepts to the world, which allows the use of those concepts to be more than just a "frictionless spinning in a void" (McDowell, 1994). Morgan and Morrison (1999) have compellingly characterized theoretical models as partially autonomous *mediators* between theories and the world. I am claiming that scientific understanding is often *doubly* mediated; experimental systems mediate between the kinds of models they describe as instruments, and the circumstances to which scientific concepts ultimately apply. It is often only by the application of these models within the microworld of the experimental system that they come to have an intelligible application anywhere else. Moreover, in many cases, the experimental model comes first; it introduces relatively well-behaved circumstances that can be tractably modeled in other ways (e.g., by a *Drosophila* chromosome map).[17]

To understand the significance of this claim, we need to ask what "well-behaved circumstances" means here. Cartwright (1999a, pp. 49–59) has introduced similar issues by talking about mediating models in physics or economics as "blueprints for nomological machines." Nomological machines are arrangements and shielding of various components, such that their capacities reliably interact to produce regular behavior. I am expanding her conception to include not just regular behavior, but conceptually articulable behavior more generally. I nevertheless worry about the metaphors of blueprints and machines. The machine metaphor suggests an already determinate purposiveness, something the machine is a machine *for*. With purposes specified, the normative language that permeates Cartwright's discussion of nomological machines becomes straightforward: She speaks of *successful* operation, running *properly*, or arrangements that are fixed or stable *enough*. Yet where do the purposes and norms come from? That is the most basic reason to think about experimental systems as laboratory fictions mediating between theoretical models and worldly circumstances: They help articulate the norms with respect to which circumstances could be "well-behaved," and nomological machines (or experiments with them) could run "properly" or "successfully." Scientific concepts, then, both articulate and are accountable to norms of intelligibility, expressed in these notions of proper behavior and successful functioning.

For theoretical models and the concepts they employ, Cartwright and Ronald Giere (1988) have each respectively tried to regulate their normativity in terms of their empirical adequacy, or their "resemblance" to real systems. In discussing the domain of the concept of 'force', for example, Cartwright claims that

> When we have a good-fitting molecular model for the wind, and we have in our theory . . . systematic rules that assign force functions to the models, and the force functions assigned predict exactly the right

motions, then we will have good scientific reason to maintain that the wind operates via a force. (1999a, p. 28)

Giere in turn argued that theoretical models like those for a damped harmonic oscillator only directly characterize abstract entities of which the models are strictly true, whose relation to real systems is one of relevant similarity:

> The notion of *similarity* between models and real systems . . . immediately reveals—what talk about approximate truth conceals—that approximation has at least *two* dimensions: approximation in *respects*, and approximation in *degrees*. (1988, p. 106)

To answer my concerns, however, empirical adequacy or similarity comes too late. What is at issue are the relevant respects of possible resemblance, or what differences in degree are degrees *of*. Such matters could be taken for granted in mechanics, which serves as the proximate example for both Cartwright's and Giere's discussions, because the relevant experimental systems have long been established and stabilized, in mutual adjustment with the relevant idealized models. The phenomena that constitute the domain of mechanics, such as pendula, springs, free-falling objects, and planetary trajectories, along with their conceptual characterization, could already be presumed.

Such is not the case when scientists begin to formulate and explore a new domain of phenomena. Mendelian ratios of inheritance obviously predated Morgan's work, for example, but spatialized "linkages" between heritable traits were novel. The discovery that the white-eyed mutation was a "sex-linked" trait undoubtedly provided an anchoring point within the emerging field of mutations, much as the freezing and boiling points of water helped anchor the field of temperature differences. Yet as Hasok Chang (2004, chap. 1) has shown, these initially "familiar" phenomena could not be taken for granted; in order to serve as anchors for a conceptual space of temperature differences, the phenomena of boiling and freezing required canonical specification. Such specification required practical mastery of techniques and circumstances as much or more than explicit definition.[18] Indeed, my point is that practical and verbal articulation of the phenomena had to proceed together. Likewise with the development and refinement of the instruments through which such phenomena could become manifest, such as thermometers for temperature differences or breeding stocks for trait-linkages.

Experimental systems can function in this domain-opening, concept-defining way, without having to be typical or representative of the features they exemplify. Consider again *Drosophila melanogaster* as an experimental organism. As the preeminent model system for classical genetics, *Drosophila* was highly atypical; as a human commensal, it is relatively cosmopolitan and genetically less diversified than alternative model organisms. More important, however, for *D. melanogaster* to function as a model system, its atypical features had to be artificially enhanced, removing much of its residual "natural" genetic diversity from experimental breeding stocks (Kohler, 1994, chaps. 1, 3, 8). *Drosophila* is even more anomalous in its later incarnation as a model system for evolutionary-developmental biology. *Drosophila* is now the textbook model for the development, developmental genetics, and evolution of animal body plans (Carroll et al., 2001, esp. chaps. 2–4), and yet the long syncytial stage of *Drosophila* development is extraordinary even among arthropods.

One might object that conceiving these possibly anomalous yet domain-constituting experimental systems as fictions renders the constitution of these domains merely stipulative, and thus not answerable to empirical findings. Do they merely institute normative standards for the application of a concept that are not accountable to any further normative considerations? Chang's (2004) study of the practices of thermometry shows one important reason why such norms are not merely stipulative. There are many ways of producing regular and reliable correlates to increases in heat under various circumstances. Much work went into developing mercury, alcohol, or air thermometers (not to mention their analogues in circumstances too hot or cold for these canonical thermometers to register). Yet it would be insufficient merely to establish a reliable, reproducible system for identifying degrees of heat (or cold) by correlating them with the thermal expansion or contraction of some canonical substance. Rather, the substantial variations in measurement among different putatively standard systems suggested a norm of temperature independent of any particular measure, however systematic and reproducible. Such a norm, if it could be coherently articulated, would introduce order into these variations by establishing a standard by which their own correctness could be assessed. That the development of a standard is itself normatively accountable is clear from the possibility of failure: Perhaps there would turn out to be no coherent, systematic way to correlate the thermal expansion of different substances within a single temperature scale.

Chang (2004, pp. 59–60) identifies the deeper issue here as "the problem of nomic measurement": Identifying some concept X (e.g., temperature) by some other phenomenon Y (e.g., thermal expansion of a canonical substance) presupposes what is supposed to be discovered empirically, namely, the form of the functional relation between X and Y. But the more basic underlying issue is not the identification of the *correct* functional relation, but the projection of a concept to be right or wrong about in the first place. The issue therefore does not apply only to quantitative measurement. It affects nonquantitative concepts like the relative location of genes on chromosomes or the identification of functionally significant components of cells as much as it applies to measurable quantities like temperature or electrical resistance.[19]

The most dramatic display of the normativity of experimental domain constitution, however, comes when domain-constituting systems are abandoned or transformed by a constitutive failure, or a reconceptualization of the domain. Consider the abandonment in the 1950s of the *Paramecium* system as a model organism for microbial genetics.[20] *Paramecium* was dealt a double blow. Its distinctive advantages for the study of cytoplasmic inheritance became moot when the significance of supposed differences between nuclear and cytoplasmic inheritance dissolved. More important from my perspective, however, is the biochemical reconceptualization of genes through the study of auxotrophic mutants in organisms that could grow on a variable nutrient medium. Despite extensive effort, *Paramecium* would not grow on a biochemically characterizable medium, and hence could not display auxotrophic mutations. In Elgin's terms, the cytogenetic patterns in *Paramecium* could now only instantiate, but could no longer exemplify, the newly distinctive manifestations of genes.

A different kind of failure occurs when the "atypical" features of an experimental system become barriers to adequate conceptual articulation. For example, the very standardization of genetic background that made the *D. melanogaster* system the exemplary embodiment of chromosomal genetics blocked any display of population-genetic variations and its significance for evolutionary genetics; Dobzhansky had to adapt the techniques of *Drosophila* genetics to *D. pseudoobscura* in order to manifest the genetic diversity of natural populations (Kohler, 1994, chap. 8). More recently, Jessica Bolker (1995) has suggested that the very features that recommended the standard model organisms as laboratory models of biological development may be systematically misleading. Laboratory work encourages using organisms with rapid development and short generations; these features in turn correlate with embryonic pre-patterning and developmental canalization. The choice of experimental systems thereby *materially* conceives development as a relatively self-contained process. The reconceptualization of development as ecologically mediated may therefore require exemplification in different experimental practices, which will likely employ different organisms.

4. CONCLUSION

I have been considering experimental systems as materialized fictional "worlds" that are integral to scientific conceptualization. The experimental practices that establish and work with such systems do not just exemplify conceptualizable features of the world, but help constitute the fields of possible judgment and the conceptual norms that allow those features to show themselves intelligibly.

The sense in which these systems are "fictional" is twofold. First, they are simplified and rearranged as "well-behaved" circumstances that make some features of the world more readily manifest. The world as we find it is often unruly and unarticulated. By arranging and maintaining more clearly articulated and manifest differentiations, we create conditions for conceptual understanding that can then be applied to more complicated or opaque circumstances.[21] Second, these artificially regulated circumstances and the differentiations they manifest are sufficiently interconnected to allow the differences that they articulate to be systematically interconnected. The resulting interconnections, if they can be coherently sustained and applied to new contexts and new issues, demarcate norms for conceptual intelligibility rather than merely isolated, contingent correlations. What might otherwise be merely localized empirical curiosities instead become scientifically significant because they allow those situations to manifest an intelligibly interconnected domain of conceptual relationships. By articulating relevant conceptual norms, experimental systems join theoretical models in doubly mediating between scientific theory and the world it thereby makes comprehensible.

To talk about laboratory "fictions" in this way does not challenge or compromise a scientific commitment to truth. On the contrary, these fictional constructions help establish norms according to which new truths can be expressed (and, correspondingly, erroneous understanding can be recognized as such). Moreover, as I have argued, such fictional constitution of norms is not stipulative or voluntaristic, but is instead itself normatively accountable to the world. Conceptual articulation through experimental practice is vulnerable to empirical failure, but such failures are manifest in conceptual confusion or incoherence rather than falsehood; in Hacking's (1984) terms, they challenge the "truth-or-falsity" of claims made using those concepts, rather than their truth. Haugeland (1998, chap. 12) has rightly emphasized that such normative accountability in science requires both resilient and reliable skill (that allows scientists to cope with apparent violations of the resulting conceptual norms, by showing how they are *merely* apparent violations), and a resolute willingness to revise or abandon those norms if such empirical conflicts cannot be adequately resolved. Yet these crucial skills and attitudes cannot come into play unless and until there is a conceptually articulated domain of phenomena toward which scientists can be resilient and resolute. The discursive function of laboratory fictions helps open and sustain such conceptual domains, and thereby helps make the attainment of scientific truth and the recognition of error possible.

NOTES

1. Cartwright (1999a) now recognizes that what was at issue in her concerns about whether and how laws accurately describe actual circumstances is not their truth, but their meaning (she interprets the relevant aspect of their meaning to be the *scope* of their application). Her revised view differs from

mine in at least two crucial respects, however. First, her account is one-dimensional rather than two-dimensional: She determines the scope of laws based upon only the circumstances of their application rather than the connection between circumstances and consequences. Second, she thinks that the only relevant circumstantial criterion is the empirical adequacy of the models that could connect law and circumstances, whereas I think empirical adequacy is only one among multiple relevant considerations. See Rouse (2002, pp. 319–334).

2. I adapt this two-dimensional account of conceptual meaning from Brandom (1994).

3. Humphreys (1994), Hughes (1999), Norton and Suppe (2001), and Winsberg (2003).

4. Although I will not belabor the point here, it is relevant to my subsequent treatment of experimental systems as "laboratory fictions" that, strictly speaking, the Michelson/Morley experiment does not instantiate the constant velocity of light in different inertial frames, because the experiment is conducted in a gravitationally accelerated rather than an inertial setting.

5. Aristotle, *Metaphysics* IX, chap. 6, 1048a. Hacking's initial discussion of the creation of phenomena criticized just this conception of phenomena as implicit or potential components of more complex circumstances:

> We tend to feel [that] the phenomena revealed in the laboratory are part of God's handiwork, waiting to be discovered. Such an attitude is natural from a theory-dominated philosophy. . . . Since our theories aim at what has always been true of the universe—God wrote the laws in His Book, before the beginning—it follows that the phenomena have always been there, waiting to be discovered. I suggest, in contrast, that the Hall effect does not exist outside of certain kinds of apparatus. . . . The effect, at least in a pure state, can only be embodied by such devices. (Hacking, 1983, pp. 225–226)

6. I have developed these criticisms of Davidson and McDowell more extensively in Rouse (2002).

7. Davidson himself would argue that the "triangulation" involved in the interpretation and articulation of concepts is not merely intralinguistic. Yet he also sharply distinguishes the merely causal prompting of a belief from its rational, discursive interpretation and justification. For discussion of why his view commits him despite himself to understanding conceptual articulation as intra-linguistic, see Rouse (2002, chaps. 2 and 6).

8. For a useful discussion of how philosophical conceptions of literary language need to take into account the conceptions of discourse that now guide contemporary literary theory, see Bono (1990).

9. Notable defenses of the systematic character of experimental systems and traditions include Rheinberger (1997), Galison (1987, 1997), Klein (2003), Kohler (1994), and Chang (2004).

10. Fleck, at least, was not unaware of the role of experimental systems in conceptual articulation, although he did not quite put it in those terms. One theme of his study of the Wassermann reaction was its connection to earlier vague conceptions of "syphilitic blood," both in guiding the subsequent development of the reaction and also thereby articulating more precisely the conceptual relations between syphilis and blood. He did not, however, explicitly connect the systematicity of experimental practice with its conceptual-articulative role. Hacking was likewise also often concerned with conceptual articulation (especially in the papers collected in Hacking [2002] and Hacking [1999]), but this concern was noticeably less evident in his discussions of laboratory science (e.g., Hacking, 1983, chaps. 12, 16; 1992).

11. One might go further, in the spirit of McDowell's (1994) criticism of David-son. On such an analogous line of criticism, Hacking's account of self-vin-dication would refute skeptical doubts about the epistemic justification of experimental science at the cost of losing one's grasp of the semantic con-tentfulness of the concepts deployed in those self-vindicating domains. If this criticism were correct, then Hacking "manages to be comfortable with his coherentism, which dispenses with rational constraint on thinking from out-side it, only because he does not see that emptiness is the threat" (McDowell, 1994, p. 68).

12. John Haugeland (1999, chap. 13) suggests the locution "letting be" to expli-cate what we both mean by the term 'constitution.' Constitution of a domain of phenomena ("letting it be") must be distinguished from the alternative extremes of creation (e.g., by simply *stipulating* success conditions), and merely taking antecedently intelligible phenomena to "count as" something else.

13. My emphasis upon inferential articulation as the definitive feature of con-ceptualization is strongly influenced by Brandom (1994, 2000, 2002), with some important critical adjustments (see Rouse, 2002, chaps. 5–7). Inferen-tial articulation is not equivalent to linguistic expression, because conceptual distinctions can function implicitly in practice without being explicitly artic-ulated in words at all. Scientific work is normally sufficiently self-conscious that most important conceptual distinctions are eventually marked linguisti-cally. Yet the central point of this chapter is to argue that the inferential artic-ulation of scientific concepts must *incorporate* the systematic development of a domain of phenomena within which objects can manifest the appropriate conceptual differences. The experimental practices that open such a domain thereby make it possible to form judgments about entities and features within that domain, but the practices themselves already articulate "judgeable con-tents" prior to the explicit articulation of judgments. My commitment to semantic inferentialism has some important affinities to Suárez (this volume), but also one very important difference, I think. Suárez emphasizes the infer-ential *uses* of scientific representations, including fictional representations. I emphasize instead that representational relations are themselves inferen-tially constituted, and I go even beyond Brandom in emphasizing the ways in which causal interaction with the world in experimentation is caught up within the discursive, inferential articulation of conceptual understanding.

14. The distinctively revealing character of negative descriptions was brought home to me by Hanna and Harrison (2004, chap. 10).

15. Classical-genetic loci within any single chromosome are especially cogent illustrations of my larger line of argument, because prior to the achievement of DNA sequencing, any given location was only identifiable by its relations to other loci on the same chromosome. The location of a gene was relative to a field of other genetic loci, which are in turn only given as relative locations.

16. Hanna and Harrison sometimes seem to contrast nomothetic objects, of which chess pieces and name-bearerships are prototypical, to natural objects. By coming to recognize the "nomothetic" character of scientific domains, my argument shows the coincidence of the domains of their "nomothetic objects" and objects more generally.

17. Marcel Weber (2007) argues that the classical gene concept as it was mate-rially articulated in *Drosophila* chromosome maps played a crucial role in articulating the molecular gene concept that is often taken to have supplanted it. The extensively interconnected maps available for *Drosophila* were a vital resource allowing the identification of the relevant regions of the chromo-some from which the coding and regulatory sequences could be isolated. As

Weber succinctly concludes, "the classical *gene* may have ceased to exist, but the classical gene *concept* and the associated experimental techniques and operational criteria proved instrumental for the identification of molecular genes" (2007, p. 38).

18. Chang (2004) argues that in the case of state changes in water, ironically, ordinary "impurities" such as dust or dissolved air, and surface irregularities in its containers, helped *maintain* the constancy of boiling or freezing points; removing the impurities and cleaning the contact surfaces allowed water to be "supercooled" or "superheated." My point still holds, however, that canonical circumstances needed to be defined in order to specify the relevant concept, in this case temperature.

19. One might argue that chromosomal locations in classical genetics were quantitative properties also, either because they were assessed statistically by correlations in genetic crossing over, or because spatial location is itself quantitatively articulable. Yet crossover correlations were only measures of relative location, and because locations were only identifiable through internal relations on a specific chromosome map, I would argue that these were not yet quantitative concepts either.

20. For detailed discussion, see Nanney (1983). Sapp (1987) sets this episode in the larger context of debates over cytoplasmic inheritance.

21. It is no part of my claim to suggest that experimental practices stand alone in this role. Mark Wilson's (2005) magnificent book *Wandering Significance*, for example, shows in exquisite detail how mathematics often plays an indispensable role in further articulating concepts beyond the range of their initially specified applications, often in rather patchwork ways.

4 Models as Fictions

Anouk Barberousse and Pascal Ludwig

1. INTRODUCTION

The words "scientific model" refer to such a variety of entities that it is difficult to say anything that would be true about *all* kinds of scientific models. For instance, not all of them are abstract entities, because the wood models of molecules, and their contemporary surrogates, namely, three-dimensional computer-generated images, are concrete models, the interest of which being that they can be easily handled and looked at from different points of view. Likewise, models are not all mathematically presented, even if many of them are.

However, we claim that all scientific models have some properties in common: All of them are representations, in the sense that they stand for something else (namely, the system they are used to investigate), and possess intentional content. Representations may be linguistic (in that case, they are descriptions), partially mathematical, diagrammatic, pictorial, or even in three dimensions (3D) (anything can be a representation, actually). Even this claim is not uncontroversial, because some models are mostly used as predictive tools whose representational target is hard to tell. Some models in finance, for example, are used as mathematical black boxes whose representational relation to the situation at hand is unclear. However, we maintain that scientific models do have a representational content, even when it is not easily recovered from their predictive capacities.

Claiming that models are representations amounts to putting them in the same category as theories. The distinction between models and theories is not straightforward. For instance, physicists usually say that today's "standard model" of elementary particles is our best available *theory* of microphysics; in the 1913 paper presenting the model of the hydrogen atom today bearing his name, Niels Bohr equally uses the words "model" and "theory" to describe it. Usage is not completely fixed. One of the aims of this chapter is to explain why.

Even when the black-box use of mathematical models is left out, the claim that models are representations is not straightforward, for the nature of their representational target is problematic: What are models representations of?

Some aspects of models may be literally false about the observational or experimental situation in which they are used and that they contribute to investigate and better understand. If models do not truthfully represent the situation under investigation, what could their representational content be? We insist that to say that models may be literally false about their observational or experimental target-situation cannot be the whole story: We need to identify their *usable* content, not primarily what is false about them.

Our main thesis is that models represent fictional situations, namely, situations that are not, and cannot (for all we know) be, instantiated in our world. In other words, models are *fictions*, that is, precisely, representations of fictional situations.[1] Moreover, representing fictional situations gives them the power to convey new usable scientific knowledge. The rest of the chapter is devoted to clarifying and justifying these two claims. We place this effort along the same line as Arthur Fine's defense of fictionalism (for instance in chap. 2 of this volume). Our aim is indeed to give a precise, renewed meaning to this position in philosophy of science. Fictionalism often lacks a precise definition of its most important notion, namely, fiction, and an analysis of the relationships between fiction, imagination, and scientific practice. We provide such an analysis in a representationalist framework.

We first make explicit what fictions (or representations of fictional situations) are, and how human beings process them cognitively. We do that in section 2 and apply this conception of fiction to models in section 3. In section 4, we present two examples. In section 5, we show how our analysis of models and modeling allows for a typology of scientific models, in spite of their diversity.

2. FICTIONS

Generally speaking, the set of all representations[2] may be divided into two broad classes: the representations that directly represent the circumstances in which they are produced on the one hand, or that are used to claim truth for certain propositions, and the representations that do not, on the other hand. The former convey propositions that can be directly evaluated as true or false in these circumstances. The notion of direct representation, although intuitive, may be explicated as follows:

> **Direct representation** *A* directly represents *B* if any interpreter of *A* can understand by herself that *A* represents *B*, *if* she is sufficiently familiar with the style of representation instantiated by *A*, where a style of representation is the set of all conventions used in the production of *A*.

For instance, a report on the nuclear industry in Iran, a financial page, and a description of the Louvre in Baedeker's guide to Paris, are direct representations; their authors intend to provide their readers with faithful information about what they represent. Moreover, it is a major feature of direct representations that they can, and should, be evaluated as conveying faithful information about what they represent. Anyone reading a financial page, for instance, immediately understands that it gives her information about the stock exchange. On the other hand, Marcel Proust's *A la recherche du temps perdu*, Martin Scorcese's *Casino*, and Picasso's *Guernica* do not directly represent the circumstances in which they were produced. They clearly have other (representative, expressive, etc.) aims, and their interpreters immediately understand that they do so.[3] These representations are fictions.

Most theorists of fiction consider that the intentional content understood by the interpreter of a fiction is propositional, and may be expressed by sentences in a natural language. The main justification in favor of this analysis is that attributions of imaginative attitudes may indeed by expressed by sentences like "The reader imagines *that Swann is in love with Odette*." However, this analysis is misleading. According to the current, Kripkean semantics for proper names, the phrase "Swann is in love with Odette" does not express any proposition, because neither "Swann" or "Odette" has a referent in our world. Moreover, according to the propositionalist account, the phrase "George W. Bush and Hugo Chavez are one and the same person" in the sentence "John imagines that George W. Bush and Hugo Chavez are one and the same person" which could describe an imaginative project aroused by a novel, expresses a contradiction, because there exists no possible world in which George W. Bush and Hugo Chavez are one and the same person.

Ad hoc descriptivist accounts of the referents of proper names within fictions have been proposed (cf. Lycan, 1994). They have two main drawbacks. First, they are ad hoc, which implies that the class of proper names is not conceived as a unified one, contrary to our linguistic intuitions. Second, these descriptivist accounts cannot be extended to the referents of other names, phrases, or other symbols. Because of this lack of generality, we move to another analysis of fiction.

Following Currie and Walton, we claim that defined broadly, the function of a fiction is to make its interpreters imagine a certain intentional content, or to make believe that some proposition is true[4] (cf. Currie, 1985, p. 387; Currie, 1990; Walton, 1990, pp. 13–21; Currie & Ravenscroft, 2002, chap. 1). Make-believe, as a pervasive element of human experience, should not be conceived as *opposed* to belief or to truth: it is a powerful means of achieving understanding of various situations. As Currie and Ravenscroft emphasize, it enables us "to project ourselves into another situation and to see, or think about, the world from another perspective" (2002, p. 1). "Think about the world" here is not just a figure of speech. Because imagination preserves the inferential patterns of belief (Currie & Ravenscroft, 2002, p. 13), we may *reason* about what we imagine and draw conclusions, exactly in the same way as we do with our beliefs. Moreover, because imagining "involves more than just entertaining or considering or having in mind the propositions imagined,"[5] and is rather "*doing* something *with* a proposition one has in mind" (Walton, 1990, p. 20), imaginings are constrained rather than freestanding. As Walton emphasizes, some imaginings are appropriate in certain contexts, and others not. This property makes imaginings distinctively capable of promoting cognitive ends.

In the context of Walton's theory of imagination and fiction, also adopted and developed by other scholars like Currie and Yablo, the content of a fiction may be characterized as follows:

> **Fiction:** The intentional content of a fiction is a context property, namely, the property instantiated by every context or situation that an interpreter of the fiction can imagine she is in when she correctly understands the fiction.[6]

Even if this account has to be complemented with an analysis of what is to correctly understand a fiction, it avoids the difficulties induced by propositional accounts. An interpreter can imagine she is in a context in which a proper name, or any symbol, which is empty in the actual context, possesses a referent. More precisely, the fiction prompts her to imagine that the name or symbol has a referent, because the fiction's intentional content involves the fact that in the contexts it refers to, this name or symbol is not empty. This implies that the contexts represented by a fiction have symbols *and their meanings* within the set of their components. True, most symbols must keep their usual meanings in order for the fiction to be understandable. However, new symbols may be introduced, as well as new meanings for existing symbols.

Let us examine the consequences of this account. First, it does *not* imply that fictions express contradictions. This is a major benefit, because in most cases, fictions do not express explicit, internal contradictions. If they did, how could we learn from fictions? How could they be informative? Fictions nonetheless can convey useful information and understanding: They have long been used as major pedagogical instruments, and perhaps still are. So instead of including internal contradictions, fictions rather convey contents that contradict our best theories about physical, biological, psychological, and also metaphysical matters. However, these very theories must be understood, and partly accepted, in order to understand the fiction.[7] The interplay between these two aspects is an essential component of fictions: A fiction is always grounded on at least some accepted conceptions about physics, biology, the behaviors of human beings, etc., but may deviate from pure (historical) facts, and even, in some cases, from the very laws governing physics, biology, or human beings.

Let us analyze the most widely discussed example in this context, namely, H. G. Wells's *The Time Machine*. A careful reading indicates that there is no patent contradiction in this novel. Contradictions only arise when inferences are made from the story and are confronted with our usual physical and metaphysical theories. For instance, it can be said about the same character that he was alive in 1935, and, in the future tense, that he cannot be alive in 1935 if he kills his father after having gone back in time. To account for this example, we claim that *The Time Machine* implies that the fundamental concepts of the theories that are true in the novel (like the concept of time) do not have their usual meanings: They have to be interpreted in such a way as to dismiss the contradictions arising from the application of our theories, with their usual interpretations, to the situations described in the fiction. The notion of fiction indeed involves that different (physical, metaphysical, biological, psychological) theories than our usual ones may have been developed in the different contexts it specifies, and that symbols used in these theories may have different denotations than in the actual context. These deviating theories are usually left indeterminate, as well as the meanings of their associated symbols. Generally speaking, indeterminacy is a major feature of imaginings: As Yablo (1993, sec. 10) indicates, propositional imaginings are of more or less determinate situations, or more precisely of fully determinate situations whose determinate properties are left more of less unspecified. Some of these properties may indeed be governed by deviating theories of which we do not exactly know what constraints they impose on the imagined situation.

The claim that fictions do not usually encompass plain contradictions is in keeping with the fact that we have no difficulty in interpreting a novel, for instance, as describing a consistent world. In reading it, we constantly reinterpret the written words in order to build a consistent, fictional world in our mind. This unremitting, but enjoyable exercise is part of the pleasure we experience in reading novels. It simply consists in bracketing the concepts that may involve contradictions, namely, in suspending our use of these concepts when they lead to inconsistencies. Their meanings may remain indeterminate and subjected to the only constraint that they must enter into propositions building a consistent set.

Avoiding giving certain concepts, in certain of their occurrences, their usual meaning amounts to making the meaning of concepts depend on the context in which they are used. The context of use should be understood in a broad way, including the language containing the words that express the concepts in question, *as well as* the facts setting the conventions of the language. When we imagine the intentional content expressed by a fiction, our imaginative project does not only focus on counterfactual circumstances determined by our usual concepts, associated with the words of our usual language, considered as fixed. It may also be about the meanings of the very concepts we use in order to specify the imagined circumstances. For instance, when we imagine that we go back in time, the very concept of time can be imagined to be freed of the theoretical constraints commonly imposed on it by our physical theories in their usual interpretation.

To sum up, we claim that fictions convey intentional contents, even though they are not (directly) about our world. To imagine is to entertain thoughts that focus on contexts in which some of the propositions we take as true according to our best theories in their usual interpretation are bracketed. This implies that the interpreter of any fiction must reorganize her concepts; this is part of the pleasure we experience when we interpret fictions. The content conveyed by a fiction is expressed through idiosyncratic concepts that are worked out within the fiction itself. We claim that, for all these reasons, the intentional contents conveyed by fictions may play an important role in the enterprise of knowledge production, and show how in the following section.

3. MODELS

Prima facie, models are different from fictions, because their function is chiefly epistemic. Moreover, modeling is severely constrained by experimental data and the relevant set of accepted background theories. The imaginative game seems considerably limited. However, as soon as it is acknowledged that it is uninteresting to emphasize that models are sometimes literally false representations, their role in the pursuit of scientific knowledge is in need of explanation. We suggest that the framework we put forward in the analysis of fiction may also account for the use of scientific models.[8] More precisely, we think that a common analysis of scientific models and fictions may shed light on a vast branch of human cognitive activity.

3.1 Models and Theories

Most models entertain intricate and often conflicting relationships with accepted theories.[9] Even when a model deliberately conflicts with the current theories applying to the same phenomena, these theories are necessary components of its scientific utility, because they make up its semantic and epistemic background. No model stands by itself, so to say; it backs onto vast amounts of theoretical and empirical knowledge, which provide it with concepts, more or less confirmed hypotheses, and, in some cases, physico-mathematical representational tools like equations and their typical solutions or "templates" (Humphreys, 2004, pp. 60–67). Background theories and empirical data give their meaning to the various symbols used in models. These symbols may be linguistic, algebraic, diagrammatic, or pictorial.

It is thus not surprising that it is sometimes difficult to tell the difference between a model and a theory, because a model and a theory applying to the same domain of phenomena are made up of the same components. The main difference is that usually a model's generalizations do not harmoniously add up to current theoretical generalizations, but rather contradict them.

Redhead, among others, mentions that at least some models, namely, those models whose aim is to make mathematical computations possible, logically contradict the theories that generate them (cf. Redhead, 1980). However, he does not emphasize how surprising this may appear. Because the background theories are the best available at a time, any representation contradicting them is prima facie likely to be less satisfactory than they are.

However, models are, in some important respects, more useful and therefore more satisfactory representations than theories: They allow for predictions that would be impossible to work out otherwise, or for looking into details of phenomena that are out of the reach of their background theories. The models Redhead refers to, for instance, are indispensable to apply the relevant theories to phenomena. These theories are otherwise mute about the concrete situations scientists seek to better understand. We claim that the tension between models and their background theories calls for a cognitive as well as a semantic explanation. If it is not given, one has to conceive of models as useless representations, conveying no usable information, which is highly paradoxical and contradicts all that is known about the use of models in every branch of scientific activity.

Let us examine a possible objection against the preceding presentation of the semantic relationships between models and their background theories. One could say that the mentioned difficulty is illusive, because within the model, the generalizations of the background theories simply do not apply; thus, no contradiction follows. However, we have to pay attention to the fact that the sentence at stake, namely, (*) "Within the model, the generalizations of the background theories do not apply," can be interpreted in at least three different ways.

1. A first interpretation is that (*) means that the model represents a physically (or chemically, or biologically, or psychologically, etc.) possible situation in which the involved theoretical generalizations do not hold. This first interpretation is simply inconsistent. When using the billiard-ball model of a gas in order to predict when it will reach equilibrium from such and such out-of-equilibrium state, a physicist by no means claims that quantum laws do not hold in some physically possible sample of some gas. The very notion of a physically possible sample of a gas necessarily implies that it is subjected to quantum laws.
2. A second, radical interpretation of (*) amounts to giving up the concept of a model. After all, scientific life is full of competing and contradictory hypotheses. In this interpretation, models are just new theories proposing new hypotheses competing with the currently accepted ones. Contrary to the preceding one, this interpretation is consistent, but empirically inadequate. It is true that *some* models are incipient theories, but not all of them are. The huge number of models that are produced with the "sole" aim of computing the consequences of theories are obvious counterexamples. (This aim is not as pedestrian as the

expression "the sole aim" indicates. It usually requires a lot of invention [see for instance Morgan & Morrison, 1999].) To put it another way, most models do not claim to contain new hypotheses, competing with the currently accepted ones. They do not pretend to play the same explanatory game, which would imply fitting into a (partly) unified explanatory picture. They have more limited and varied goals.

3. The third interpretation of (*) is just an application of our analysis of fiction. One can understand (*) as saying that (a) one can *imagine* that within the (fictional) context represented by the model, the generalizations of the background theories do not have their current meaning, and (b) their meaning is subjected to the constraint that they should be consistent with the model's generalizations.

The scientists using the model or evaluating its scientific adequacy are pressed to imagine that the concepts entering the relevant generalizations of the background theories have new meanings that are partly indeterminate. Just as we overstep, in interpreting a fiction, the limits of natural language, for instance, in attributing meanings to proper names that are empty in the current language, in interpreting a scientific model we overstep the constraints imposed by the relevant theoretical language. As Max Black said, "Using theoretical models consists in introducing a new language or dialect" (1962, p. 229).

The indeterminacy of certain theoretical concepts in scientific models is a seemingly counterintuitive consequence of our approach. However, it happens that the exact extension of a concept is not so important. Consider again classical models of gases, in which quantum laws are bracketed. In such models, the concept of a molecule does not have its current meaning, but its precise denotation is irrelevant; only certain properties of these molecules* are important in the model, like their interaction function and their mass. Their detailed inner structure is generally irrelevant. Molecules* in classical models of gases are thus sketches of material entities, of which only certain properties are explicitly considered. The specific generalizations governing them (namely, the application of Newton's laws to their interaction function) gives a definite structure to the space of conceptual possibilities that models are designed to explore. It happens that physicists wish to complete the first sketch, and consider more properties, for instance, the relative motion of atoms within a molecule*. This gives rise to new models.

3.2 Models of Data

Our analysis also applies to models of data (Suppes, 1974; van Fraassen, 1985). A prima facie objection is that the interest of exploring the space of conceptual possibility related to the representation of a set of empirical data is not immediately obvious. Nevertheless, scientists do so when the available theoretical representation is badly incomplete and when there is no other way of completing it than by imagining how it could be completed. The hope guiding this enterprise is that new empirical data will further constrain the conceptual exploration. We come back to models of data in section 5.2.

4. EXAMPLES

4.1 The Ehrenfest Model

Our first example is a model aiming at a better understanding of the phenomenon of thermalization. The phenomenon of thermalization can be presented by the following, simplified case. Let us consider two bodies A and B at different temperatures. Let us put them in thermal contact and isolate them from the surrounding. Thermodynamics predicts that the difference between A's temperature and B's temperature will tend toward zero: this is what it is called "thermalization." The heat, so to speak, flows exclusively from the warmer body to the colder one.

We will see how a particularly rough model of matter called the "Ehrenfest model" sheds theoretical light on this phenomenon by bringing into relief the few characteristics of more sophisticated mathematical models of gases that are actually responsible for thermalization. This crude model is surely not the last word on this question, because the further question of why these few characteristics obtain in gases is left unanswered within it. However, it helps distinguishing what in the more sophisticated models explains thermalization.

The question of the explanatory responsibility of the various elements of a mathematical model is not much discussed; however, it is all the more difficult to answer because the mathematics involved is intricate. The parts of an equation seldom bear their physical meanings on their sleeves. They have to be inferred through shaky and even uncertain inferences. As statistical mechanical models are notoriously intricate mathematical objects, it is usually uneasy to "read" the distribution of the explanatory charge on its mathematics. The Ehrenfest model frees itself from sophisticated equations and highlights which physical properties explain thermalization.

Thermalization is theoretically described by thermodynamics, which is a phenomenological theory in the physicists' sense. It consists in relations among measurable quantities and does not postulate any unobservable entities. Its phenomenological relations are explained by another theory, statistical mechanics. Statistical mechanics derives thermodynamic and hydrodynamic laws from dynamical and statistical hypotheses about the motion of molecules. In the case of two bodies put into thermal contact, its prediction is slightly different from the thermodynamical prediction: It predicts that the temperatures' difference will exponentially tend toward zero *with a very high probability*. Statistical mechanics does not exclude that the warmer body becomes still warmer, and the colder still colder, but attributes a very weak probability to such an event.

The Ehrenfest model, although a very crude model of molecules, allows for a nice explanation of the difference between the two theories. Let us consider a box with two compartments A and B separated by a wall. Compartment A initially contains N numbered balls. Let us imagine the following process: We randomly draw a number between 1 and N, we take the corresponding ball in A, and put it into B. After some time, it will happen that the drawn number will correspond to a ball that is already in B. Then it will be put into A again. The further this game goes on, balls go from A to B, and, sometimes, from B to A. The draws are independent and all numbers between 1 and N are equiprobable. As long as there are more balls in A than in B, the probability that a ball goes from A to B is greater than the probability that a ball goes from B to A. In this case, the flow of balls is on average unidirectional, as the flow of heat is in the phenomenon of thermalization.

Probability theory gives us more information about this game. The number of balls in A is governed by a Markov chain. Its average exponentially decreases until it reaches N/2. Moreover, one can predict with probability 1 that the system will come back to its initial state, in which all the balls are in A, after 2(exp N) steps. This corresponds to a very long time, even when there are only a few hundreds balls in the system. This prediction is the counterpart of Poincaré's recurrence theorem in statistical mechanics, which predicts with probability 1 that if the two bodies in thermal interaction can be represented by conservative dynamical systems, the whole system will come back to its initial state out of equilibrium after a very, very long time.

It is clear that the Ehrenfest model represents a situation that is incompatible with our best physical theories, and particularly with quantum mechanics, or even with classical mechanics as applied to the microscopic scale (which is a sufficiently good approximation in most cases). In none of these theories is it possible to let molecules cross the wall one by one without disturbing the totality of the remaining system. This model allows us to imagine what would happen in a situation where molecules are like microscopic balls, which amounts to bracketing what we currently know about molecules. In the imagined context, the word "molecule" acquires a new meaning, because in the model we use a different concept from our usual concept of molecule.

The Ehrenfest model's epistemic utility comes from the fact that the fictional situation and the real one—as described by our best theories—share at least two important properties. In the model as well as in the real situation, the system reaches an equilibrium state after a relatively short time, and one can predict with probability 1 that it will come back to its initial state after a very long time. Of course, the reasons justifying these predictions are different: When we describe the thermalization phenomenon within statistical mechanics, we ground our predictions on the statistical properties exhibited by dynamical systems, whereas in the Ehrenfest model, we only use probability theory.

What inferences does this comparison allow for? The dynamics of the Ehrenfest model exhibits two surrogates of well-known statistical mechanical phenomena, namely, evolution toward equilibrium and recurrence; can we infer from this fact that the *processes* leading to equilibrium and to recurrence also share important properties? The answer is "yes." This is the main interest of the model. However, it is a prima facie paradoxical "yes," because statistical mechanics implies that the process leading to equilibrium and to recurrence depends on the causal, dynamical relations among molecules, whereas there are no causal relations at all among the balls in the Ehrenfest model. The conclusion we have to draw from the Ehrenfest model is that real molecules are subjected to a process that is similar to a random draw, in spite of their causal, dynamical interactions.

More precisely, the epistemic interest of the Ehrenfest model is that it allows us to isolate the probabilistic component of the statistical mechanical representation, and to show that this component alone, when separated from the causal-dynamical component, is responsible for some important aspects of the behavior of systems of molecules (or gases). Another consequence may be derived from the success of the model: When too few balls are involved (typically less than 1,000), the concepts of equilibrium and of recurrence lose their meaning. Accordingly, it can be inferred that at least three components are involved in the thermalization phenomenon: the process of random draw, or one similar to it; the number of molecules; and the causal-dynamical interactions among molecules, which enter in the explanation of other aspects of thermalization than those exhibited by the Ehrenfest model.

A last point should be made about the meaning of the expression "the process of random draw, or one similar to it," because "similar" is a notoriously vague relational predicate. A process similar to random draw here is a process that can be represented by a random draw without losing any relevant information.

4.2 Implications of Our Analysis of the Ehrenfest Model

The Ehrenfest model plays a striking anti-Duhemian role in allowing for the separation among three important hypotheses that are glued together in most applications of statistical mechanics to gases. Generally speaking, scientific models, representing fictional situations, are powerful breakers of the Duhem–Quine cement. In fictional situations, hypotheses need not be interdependent. The imaginative game is only constrained by a consistency requirement, but not by the Duhem–Quine thesis. For sure, every fictional situation generates its own Duhem–Quine constraints; however, the modeling game precisely consists in imagining other fictional situations in cases where we want to consider the consequences of different hypotheses one by one. The consistency requirement should not be underestimated: It is not so easy to build up fictional situations that exactly test the relative explanatory weights of the different hypotheses that are involved in the same theory. The Ehrenfest model is a remarkable example of such an achievement. Other similar examples are the various versions of the Ising model (Hughes, 1999, for a philosophical analysis; Vichniac, 1984).

4.3 Bohr's Model of the Atom

In 1913, Bohr built up a detailed model of the hydrogen atom that had many advantages over its predecessors, like Rutherford's model. It is stable, and its predictions strikingly correspond to available empirical data, peculiarly to Balmer's formula describing spectral rays: $f = v/4 - v/n^2$, where $n > 3$ and v is a postulated fundamental frequency. In 1913, there was no explanation at all of why Balmer's formula corresponded so well to available data.

Let us first set out which ingredients were available in 1913 that could lead to understand the inner structure of the hydrogen atom. The classical theories (namely, the prequantum theories of mechanics and electromagnetism) describe the hydrogen atom in the following way: the circular motion of the electron is periodic, the frequency of the emitted radiation is equal to the electronic motion's frequency, and harmonics may also take part in the radiation—however, no harmonic appears in Balmer's formula, which is empirically very precise.

Besides classical theories, Planck's work on black-body radiation was also available in 1913. One of its consequences is that a frequency multiplied by the constant h (called by Planck's name) gives an energy. This suggests that the equation

$$hv/4 - hv/n^2 = 0 \quad (1)$$

may be interpreted as an energy balance between some initial and some corresponding final state.

Armed with all of this, Bohr introduced three revolutionary hypotheses constituting the atomic model bearing his name.

1. *In the hydrogen atom, the atom's energy is quantified.* Energy can only assume certain (negative) values:

$$E_0 = -hv; E_1 = -hv/4; E_2 = -hv/9; \ldots; E_n = -hv/n^2 \quad (2)$$

Each E_i is the energy of one of the atom's stationary state; n is called a "quantum number" and can be any integer.

The concept of a stationary state was radically new at the time. It did not belong to any classical theory, and has thus no meaning within any of them. We shall see that the hypotheses in which it enters contradict the classical theories. It was waiting for a consistent theory, so to speak.

2. *The hydrogen atom does not radiate except when it "jumps"' from a stationary state to another.* In this case, the frequency of the emitted frequency is given by an equation of the same type as (1). The concept of "jumping from a stationary state to another" is also a brand new one that does not have any meaning in the available theories.

3. *The ratio between the electron's kinetic energy and its rotational frequency is an integral multiple of h/2.* This hypothesis allows one to compute the series of allowed energies given in (2), and is in remarkable agreement with empirical measurements of hv.

Neither the first hypothesis nor the third may be justified by classical electromagnetism. Moreover, nothing in classical electromagnetism suggests that it ceases to be valid inside atoms. The second hypothesis implies that an electron's radiation frequency differs from its rotational frequency, which contradicts many consequences of the classical theories. Another aspect of Bohr's model is utterly unintelligible in the context of the available theories: The emitted radiation's frequency is determined by the electron's energy before radiating and by its energy *after* having emitted a radiation, which is counterintuitive because it seems to involve backward causation.

Bohr's model of the atom can be viewed as representing a situation that is utterly unintelligible from the standpoint of its classical background theories, but that allows for incredibly precise and exact predictions. The content of the model, that is, a set of contexts in which electrons jump from stationary states to stationary states, can only be imagined and is only intelligible by someone who exactly knows the consequences of the classical theories and Planck's work on black-body radiation. Its main epistemic justification is that it allows for very good predictions. From our retrospective point of view, we can say that Bohr described the hydrogen atom with concepts belonging to a *future* theory: He represented a fictional hydrogen atom with concepts the meaning of which was not available yet at the time.

5. TOWARD A TYPOLOGY OF SCIENTIFIC MODELS

It has often been said that models are too heterogeneous entities to be submitted to a rational typology (Morrison & Morgan, 1999). We suggest that our analysis may provide a guiding principle for such a typology. This principle can be derived from Ramsey's map metaphor, according to which a system of beliefs is analogous to a map of the world that allows one to find one's bearings. Scientific theories contribute to such a map as far as they can be objects of belief. The map provided by scientific theories is, however, blatantly incomplete. Unexplored areas are many. Moreover, we do not know how to relate the supposedly well-known areas. For instance, in certain domains, we may possess a vast amount of empirical data on the one hand, and theoretical principles on the other hand, without being able to relate data to theoretical principles. When scientists do not *know* how to express hypotheses connecting known areas, they *imagine* situations in which the connection is made. From this imaginative activity arise two types of models, according to the object of the imaginative project: data or theoretical principles. A third type of model allows for testing a hypothesis independently of the set to which it belongs; it is once again imagination that permits breaking up the Duhemian cement among hypotheses.

5.1 Prospective Models

In a situation of epistemic incompleteness in which the connection between data and theories is hardly made, a first strategy is to imagine a context, that is, to build up a model, in which: (a) some hypotheses subsume the available empirical data, and (b) these hypotheses are deducible from theoretical principles. This type of model may be called "prospective" because it arises when a scientist projects herself, so to speak, into a future time where the hypotheses will be justified by theoretical reasons. Bohr's model of the atom is an example of a prospective model in this sense.

How does a prospective model differ from a (still incomplete) theory, or from a theory-to-come? The building up of a theory, however incomplete or provisional, is submitted to an important constraint that does not apply to prospective models, namely: A (provisional) theory has to entertain determinate relations with other available theories. Competition is an obvious example of such relations, as well as limiting relations and reduction relations. To put it briefly, a new, provisional theory T may compete with T_1, reduce T_2, and have T_3 as a limiting case. On the other hand, prospective models do not have to entertain such relations with available theories. They usually cannot purport to compete with any such theory because they lack sufficient explanatory power.

The three constitutive hypotheses of Bohr's model cannot be seen as competing with classical mechanics or classical electromagnetism. They barely contradict the consequences of these theories. Moreover, Bohr's hypotheses appeared in 1913 as isolated elements, miraculously accounting for experimental data but otherwise devoid of any justification. They would not be integrated in any genuine theory until 1925–1926, when a solid theory of the microscopic domain has been developed.

Prospective models allow for the exploration of new areas and for the invention of new connections between already explored areas in the following way. From a successful prospective model, the inference can be drawn that in a situation (still purely imaginary) in which the hypotheses of the model would be justified by overarching theoretical principles, the empirical data that are still surprising or just weird could be fully explained. Therefore, a successful prospective model urges one to modify the theory used as a starting point (classical mechanics and electrodynamics in Bohr's case) in such a way as to integrate it.

5.2 Bridge Models (or Models of Data)

Although prospective models represent situations in which theories are different from the current ones, in bridge models, the current theories are maintained but data are supposed to be different from what we know them to be. Cartwright emphasizes how frequent these models are in mathematical physics:

> The theory has a very limited stock of principles for getting from descriptions to equations, and the principles require information of a very particular kind, structured in a very particular way. The descriptions that go on the left—the descriptions that tell what there is—are chosen for their descriptive adequacy. But the 'descriptions' on the right—the descriptions that give rise to equations—must be chosen in large part for their mathematical features. This is characteristic of mathematical physics. The descriptions that best describe are generally not the ones to which equations attach. (Cartwright, 1983, p. 131)

When a physicist wants to subsume the available data under a theory, she first has to transform them to fit the available mathematical forms. This can only be done with the help of a certain amount of fiction, namely, with the help of bridge models. The physicist has to imagine that these data describe infinite planes, or mass points, instead of three-dimensional solids. This process is described by Suppes (1974) and van Fraassen (1985) as the construction of models of data. In the mass point example, the idealization resulting in the fiction is "controlled" (Sklar, 1993, p. 258). When we represent a three-dimensional solid by a mass point, we nevertheless know how to represent the action of a force on other parts of the body than its center of mass. Moreover, we know in this case that a more complete representation would be a useless sophistication because it only improves predictions in a negligible way and is computationally expensive. This is why the mass point idealization is justified in Newtonian mechanics: It makes computations much easier at a low representational cost. Other idealizations are not as easily controlled, however.

The hypotheses building up bridge models or models of data, even in the case of controlled idealizations, can hardly be justified by any available theory. Their justification is purely pragmatic. The resulting models are truly useful fictions representing imaginary situations in which the empirical data would have the right mathematical form from the start. How useful? Redhead (1980), among others, attempts to measure the distance between modeled and observed situations. However, the formal notion of distance he introduces fails at determining whether a given model of data is better than another one even when the latter is closer to the observed situation according to his measure.

5.3 Anti-Duhemian Models

The imaginative variation resulting in a model may focus on a small set of principles and concepts of a theory instead of the whole theory. In this type of model, a few principles are bracketed because they are thought as irrelevant to the data at hand, or the aspect of data the scientist wants to investigate. Such models have a testing function: If their constitutive hypotheses (which do not include the bracketed set P) can correctly subsume the data at hand, one may conclude that the set P is actually irrelevant to the explanation of the phenomenon at hand.

The Ehrenfest model is an example of such testing, or anti-Duhemian, models. The bracketed principles in this example are those describing the dynamics of molecules, as said earlier (section 4.3). When one leaves the dynamics of molecules aside and replaces it in imagination by a random drawing process, the result is in a remarkable agreement with some properties of observed phenomena (provided that the number of "molecules" is large enough). From the success of this model, one can conclude that the theoretical hypotheses bearing on molecular dynamics are not relevant to the explanation of *these* properties (but are of course relevant to the explanation of other properties involved in thermalization processes).

In some cases, it is hardly possible to design an experiment testing the validity of a hypothesis, due to technical or Duhemian-like difficulties. In such cases, models may not be enough. *Simulations* of experiments may help perform such a test. Let us investigate, as an example, the complex hypotheses one can make about the system constituted by a new plane and its pilot. Whether these hypotheses are justified or not is crucial. However, they cannot be experimentally tested. Simulations are used instead. In certain circumstances, such simulations amount to imagining that the data provided by the flight simulator, which are perceptual inputs for the pilot, are facts. In our conception, a simulated flight is thus based on a model drawing on the pilot's real behavior in a make-believe game in order to test hypotheses on what her behavior would be in a real flight.

More generally, our conception makes sense of models like boat scale models, architectural models, wood car models, etc. that do not explicitly express generalizations on any domain. Building up such models amounts to imagining that scale transformations or simplifications are irrelevant and that the data produced by these models are reliable. This is why one can test, at least in imagination, hypotheses about the target systems by observing these concrete models.

6. CONCLUSION

We have presented a new conception of models grounded on recent approaches to fictions, namely Currie's and Walton's. We have analyzed two well-known examples in order to show how our analysis sheds light on often-discussed questions about models. Finally, we have shown that our conception allows for the elaboration of a typology of models, despite their enormous diversity.

ACKNOWLEDGMENTS

We warmly thank Mauricio Suárez and Ron Giere for helpful comments on a previous version of this chapter, as well as all the other participants in this volume.

NOTES

1. This explication of the term "fiction" may be unusual; however, because no consensus is available for its meaning, we choose this one, which nicely fits to our representationalist concerns. We give further elements of the meaning of the words "fiction" and "representation" in the next section.
2. In this chapter, we do not follow Walton's (1990) use of the term "representation" and allow for nonfictional representations.
3. It may be asked why we use the standpoint of the interpreter rather than the creator of the representation in this definition. The main reason is that representations, when they are used as such, are intended to be interpreted. The interpreter's standpoint is thus essential to their very nature, as well as her capacities to understand them *as* representations. Thanks to Ron Giere for asking this question.
4. In the framework of this chapter, only propositional imaginings are considered; however, imaginings may be objectual as well (cf. Walton, 1990; Yablo, 1993; Chalmers, 2002).
5. Cf. also Chalmers (2002, section 2): "Imagining a world is not merely entertaining a description. It may involve standing is *some* relation to a detailed description of the world in question . . . but this relation is beyond entertaining or supposing."
6. Cf. Lewis (1979). Such a property can be modeled as a set of individuals or as a set of centered worlds (Quine, 1969). We borrow the following definition from Chalmers (1996, p. 60): "A centered world is an ordered pair consisting of a world and a *center* representing the viewpoint within that world of an agent using the term in question." Because an agent and a world define a context, a set of centered worlds defines a context property.
7. As Currie and Ravenscroft emphasize, our imaginings need not be consistent with anything other than a small subset of our beliefs: Imaginings are

in various ways dependent on belief. This dependence explains why fictions are so easily understood.

8. In a recent paper, Roman Frigg (in press) independently develops a version of fictionalism that is close to ours.

9. The distinction between models and accepted theories we use here is a pragmatic one: The theories are those representations that are, at a given time, accepted as reliable guides for designing experiments and constructing models. Of course, these representations also have to satisfy some conditions that are usually attributed to theories, namely, generality; however, the main difference lies in how they are apprehended.

Part III

The Explanatory Power of Fictions

5 Exemplification, Idealization, and Scientific Understanding

Catherine Z. Elgin

1. INTRODUCTION

Science, we are told, is (or at least aspires to be) a mirror of nature. It provides (or hopes to provide) complete, accurate, distortion-free representations of the way the world is. This familiar stereotype is false and misleading. It gives rise to a variety of unnecessary problems in the philosophy of science. It makes a mystery of the way scientific models function and intimates that there is something intellectually suspect about them. Models simplify and often distort. The same phenomena are sometimes represented by multiple, seemingly incongruous models. The models that scientists work with often fail to match the facts they purport to account for. These would be embarrassing admissions if models were supposed to accurately reflect the facts. But they are not. Science is not, cannot be, and ought not be, a mirror of nature. Rather, science embodies an understanding of nature. Since understanding is not mirroring, failures of mirroring are not necessarily failures of understanding. Once we appreciate the way science affords understanding, we see that the features that look like flaws under the mirroring account are actually virtues. A first step is to devise an account of scientific representations that shows how they figure in or contribute to understanding.

2. REPRESENTATION

The term 'representation' is irritatingly imprecise. Pictures represent their subjects; graphs represent the data; politicians represent their constituents; representative samples represent whatever they are samples of. We can begin to regiment by restricting attention to cases where representation is a matter of denotation. Pictures, equations, graphs, and maps represent their subjects by denoting them. They are representations *of* the things that they denote.[1] It is in this sense that scientific models represent their target systems: They denote them. But, as Bertrand Russell noted, not all denoting symbols have denotata.[2] A picture that depicts a unicorn, a map that maps Atlantis, and a graph that charts the increase in phlogiston over time are all representations, although they do not represent anything. To be a representation, a symbol need not itself denote, but it needs to be the sort of symbol that denotes. Unicorn pictures are representations then because they are animal pictures, and some animal pictures denote animals. Atlantis maps are representations because they are maps and some maps denote real locations. Phlogiston-increase graphs are representations because they are graphs and some graphs denote properties of real substances. So whether a symbol is a representation is a question of what kind of symbol it is. Following Goodman, let us distinguish between representations *of p* and *p*-representations. If *s* is a representation *of p*, then *p* exists and *s* represents *p*. But *s* may be a *p*-representation even if there is no such thing as *p*.[3] Thus, there are unicorn pictures even though there are no unicorns to depict. There is an ideal-gas description even though there is no ideal gas to describe.

Occasionally philosophers object that in the absence of unicorns, there is no basis for classifying some pictures as unicorn pictures and refusing to so classify others. Such an objection supposes that the only basis for classifying representations is by appeal to an antecedent classification of their referents. This is just false. We readily classify pictures as landscapes without any acquaintance with the real estate—if any—that they represent. I suggest that each class of *p*-representations constitutes a small genre, a genre composed of all and only representations with a common ostensible subject matter. There is then a genre of unicorn-representations and a genre of ideal-gas-representations. And we learn to classify representations as belonging to such genres as we study those representations and the fields of inquiry that devise and deploy them. This is no more mysterious than learning to recognize landscapes without comparing them to the terrain they purportedly depict.

Nor is it the case, as Suárez contends, that on Goodman's account, fictional and factual representations get entirely distinct treatments.[4] Both are *p*-representations, because they belong to denoting genres. Factual representations simply have an additional function that fictional ones lack. Besides being *p*-representations, factual representations are (or purport to be) representations *of* something.

Some representations denote their objects. Others do not. Among those that do not, some—such as phlogiston-representations—simply fail to denote. They purport to denote something, but there is no such thing. They are therefore defective. Others, such as ideal-gas-representations, are fictive. They do not purport to denote their ostensible object. So their failure to denote is no defect. We know perfectly well that there is no such animal as a unicorn, no such person as Hamlet, no such gas as the ideal gas. Nonetheless, we can provide detailed representations *as if* of each of them, argue about their characteristics, be right or wrong about what we say respecting them, and, I contend, advance understanding by means of them.

So *x* is, or is not, a representation *of y* depending on what *x* denotes. And *x* is, or is not, a *z*-representation depending on its genre. This enables us to form a more complex mode of representation in which *x* represents *y as z*. In such a representation, symbol *x* is a *z*-representation that as such denotes *y*. Caricatures are familiar examples of representation-as. Churchill is represented as a bulldog; George W. Bush is represented as a deer in the headlights. According to R. I. G. Hughes, representation-as is central to the way that models function in science.[5] This is an excellent idea. But it needs elaboration.

The problem is this: Representation-of can be achieved by fiat. We simply stipulate: Let *a* represent *b*, and *a* thereby becomes a representation of *b*. This is what we do in baptizing an individual or a kind. It is also what we do in ad hoc illustrations as when, for example, I say, 'If that chair is Widener Library, and that desk is University Hall, then that window is Emerson Hall' in helping someone to visualize the layout of Harvard Yard. So we could take any *p*-representation and stipulate that it represents any object. We might, for example, take a tree-picture and stipulate that it denotes the philosophy department. But it is doubtful that the tree-picture, as a result of our arbitrary stipulation, represents the philosophy department as a tree.

Should we say then that representation-as requires similarity? In that case, what blocks seemingly groundless and arbitrary cases of representation-as is the resemblance between the representation and the referent. I regard this as hopeless. As Goodman, Suárez, and others have argued, similarity does not establish a referential relationship.[6] Representation is an asymmetrical relation; similarity is symmetrical. Representation is irreflexive; similarity is reflexive. One might reply that this only shows that similarity is not sufficient for representation-as. Something else effects the directionality. Then it is the similarity between symbol and referent that brings it about that the referent is represented as whatever it is represented as. The problem is this: Via stipulation, we have seen, pretty much anything can represent pretty much anything else. So nothing beyond stipulation is required to bring it about that one thing represents another. But similarity is ubiquitous. For any *a* and any *b*, *a* is somehow similar to *b*. Thus if all that is required for representation-as is denotation plus similarity, then for any *a* that represents *b*, *a* represents *b* as *a*. Every case of representation turns out to be a case of representation-as. In one way or another, the philosophy department is similar to a tree-picture, but it is still hard to see how that fact, combined with the stipulation that a tree picture represents the department, could make it the case that the department is represented as a tree-picture, much less as a tree. Suppose we add that the similarity must obtain between the content of the *p*-representation and the denotation. Then for any *a*-representation and any *b*, if the *a*-representation denotes *b*, it represents *b* as *a*. But contentful representations, as well as chairs and desks, can be used in ad hoc representations such as the one I gave earlier.

If the picture on the wall represents Widener Library, and the diagram on the blackboard represents University Hall, then the chair represents Emerson Hall. This does not make the portrait on the wall represent Widener Library as the dean. Evidently, it takes more than being represented by a tree-picture to be represented as a tree. In fact, I think that some philosophy departments can be represented as trees. But to do so is not to arbitrarily stipulate that a tree-picture shall denote the department, even if we add a vague intimation that somehow or other the department is similar to a tree. The question is, what is effected by such a representation?

To explicate representation-as, Hughes discusses Sir Joshua Reynolds's painting, 'Mrs. Siddons as The Tragic Muse.' The painting denotes its subject and represents her as the tragic muse. How does it do so? It establishes Mrs. Siddons as its denotation. It might represent Mrs. Siddons, a person familiar to its original audience, in a style that that audience knows how to read. But the painted figure need not bear any close resemblance to Mrs. Siddons. We readily take her as the subject even though we have no basis for comparison. (Indeed, we even take Picasso's word about the identities of the referents of his cubist portraits, even though the figures in them seem to resemble no one on earth.) Captioning the picture as a portrait of Mrs. Siddons suffices to fix the reference. So a painting can be connected to its denotation by stipulation. The painting is a tragic-muse-picture. It is not a picture of the tragic muse, there being no such thing as the tragic muse. But it belongs to the same restricted genre as other tragic-muse-representations. To recognize it as a tragic-muse-picture is to recognize it as an instance of that genre. Similarly in scientific cases: A spring is represented as a harmonic oscillator just in case a harmonic-oscillator-representation as such denotes the spring. The harmonic-oscillator-representation involves idealization. So it is not strictly a representation of a harmonic oscillator, any more than the Reynolds is a picture of the tragic muse.

In both cases a representation that fails to denote its ostensible subject is used to denote another subject. Since denotation can be effected by stipulation, there is no difficulty in seeing how this can be done. The difficulty comes in seeing why it is worth doing. What is gained by representing Mrs. Siddons as the tragic muse, or a spring as a harmonic oscillator, or in general any existing object as something that does not in fact exist? The quick answer is that the representation affords epistemic access to features of the object that are otherwise difficult or impossible to discern. To make this out requires resort to another Goodmanian device—exemplification.

3. EXEMPLIFICATION

Let us begin with a pedestrian case. Commercial paint companies supply sample cards that instantiate the colors of the paints they sell. Of course, the cards also instantiate innumerable other properties. They are a certain size, shape, age, and weight. They are at a certain distance from the Prado. They are excellent bookmarks but poor insulators. And so on. Obviously, there is a difference between the colors and these other properties. Some of the properties the cards instantiate, such as their distance from the Prado, are matters of complete indifference. Others, such as their size and shape, facilitate but do not figure in the cards' standard function. Under their standard interpretations, the cards serve exclusively as paint samples. They are mere instances of the other properties, but telling instances of the colors. A symbol that is a telling instance of a property exemplifies that property. It points up, highlights, displays, or conveys the property. Since it both refers to and instantiates the property, it affords epistemic access to the property.[7]

Because exemplification requires instantiation as well as reference, it cannot be achieved by stipulation. Only something that is dusky rose can exemplify that color. Moreover, exemplification is selective. An exemplar can exemplify only some of its properties. It brings those properties to the fore by marginalizing, downplaying, or ignoring other properties it instantiates. It may exemplify a cluster of properties, as for example a fabric swatch exemplifies its colors, texture, pattern and weave. But it cannot exemplify all of its properties. Moreover, an exemplar is selective in how precisely it exemplifies. A single splotch of color that instantiates dusky rose, rose, and pink may exemplify any of these properties without exemplifying the others. Although the color properties it instantiates are nested, it does not exemplify every property in the nest. Exemplars are symbols that require interpretation.

Paint samples and fabric swatches belong to standardized, regimented exemplificational systems. But exemplification is not restricted to such systems. Any item can serve as an exemplar simply by being used as an example. So items that ordinarily are not symbols can come to function symbolically simply by serving as examples. Moreover, in principle, any exemplar can exemplify any property it instantiates, and any property that is instantiated can be exemplified.

But what is the case in principle is not always the case in practice. The exemplification of a particular property is not always easy to achieve, for not every instance of a property affords an effective example of it. The tail feathers of a falcon are a distinctive shade of brownish gray. Nevertheless, a paint company would be ill advised to recommend that potential customers look at a falcon's tail in order to see that color. Falcons are so rare and fly so fast and display so many more interesting properties than the color of their tail feathers, that any glimpse we get of the tail is unlikely to make the color manifest. We could not see it long enough or well enough and would be unlikely to attend to it carefully enough to decide whether it was the color we want to paint the porch. It is far better to create a lasting, readily available, easily interpretable sample of the color—one whose function is precisely to make the color manifest. Such a sample should be stable, accessible, and have no properties that distract attention from the color. Effective samples and examples are carefully contrived to bring out particular features. Factors that might otherwise predominate are omitted, bracketed, or muted. This is so, not only in commercial samples, but in examples of all kinds. Sometimes elaborate stage setting is required to bring about the exemplification of a subtle, scarce, or tightly intertwined property.

Scientific experiments are vehicles of exemplification. They do not purport to replicate what happens in the wild. Instead, they select, highlight, control, and manipulate things so that features of interest are brought to the fore and their relevant characteristics and interactions made manifest. To ascertain whether water conducts electricity, one would not attempt to create an electrical current in a local lake, stream, or bathtub. Because the liquid to be found in such places contains impurities, a current detected in such a venue might be due to the electrical properties of the impurities, not those of water. By experimenting on distilled water, scientists bring it about that the conductivity of water is exemplified.

As Nancy Cartwright has emphasized, experiments are highly artificial.[8] They are not slices of nature, but contrivances often involving unnaturally pure samples tested under unnaturally extreme conditions. The rationale for resorting to such artifices is plain. A natural case is not always an exemplary case. A pure sample that is not to be found in nature, tested under extreme conditions that do not obtain in nature, may exemplify features that obtain but are not evident in nature. So by sidelining, marginalizing, or deemphasizing confounding factors, experiments afford epistemic access to properties of interest.

But not all confounding factors are easily set aside. Sometimes properties so tightly intertwine that they cannot be prized apart. So we cannot devise an experiment that tests one in the absence of the other. This is where idealizations enter. Factors that are inseparable in fact can be separated in fiction. Even if, for example, every swinging bob is actually subject to friction, we can represent an idealized pendulum that is not. We can then use that idealization in our thinking about pendulums, and (we hope) understand the movement of swinging bobs in terms of it. The question though is how something that does not occur in nature can afford any insight into what does. Here again, it pays to look to art.

Like an experiment, a work of fiction selects and isolates, manipulating circumstances so that particular properties, patterns, and connections, as well as disparities and irregularities, are brought to the fore. It may localize and isolate factors that underlie or are interwoven into everyday life or natural events, but that are apt to pass unnoticed because other, more prominent events typically overshadow them. This is why Jane Austen maintained that 'three or four families in a country village is the very thing to work on.'[9] The relations among the three or four families are sufficiently complicated and the demands of village life sufficiently mundane that the story can exemplify something worth noting about ordinary life and the development of moral personality. By restricting her attention to three or four families, Austen in effect devises a tightly controlled thought experiment. Drastically limiting the factors that affect her protagonists enables her to elaborate in detail the consequences of the relatively few that remain.

If our interests are cognitive, though, it might seem that this detour through fiction is both unnecessary and unwise. Instead of resorting to fiction, wouldn't it be cognitively preferable to study three or four real families in a real country village? Probably not, if we want to glean the insights that Austen's novels afford. Even three or four families in a relatively isolated country village are affected by far too many factors for the social and moral trajectories that Austen's novels exemplify to be salient in their interactions. Too many forces impinge on them and too many descriptions are available for characterizing their interactions. Any such sociological study would be vulnerable to the charge that other, unexamined factors played a nonnegligible role in the interactions studied, that other forces were significant. Austen evades that worry. She omits such factors from her account and in effect asks: Suppose we leave such factors out, then what would we see? Similarly, the model pendulum omits friction and air resistance, allowing the scientist in effect to ask: Suppose we leave these out, then what would we see?

Still the question is how this is supposed to inform our understanding of reality. That Elizabeth Bennet and Mr. Darcy, who do not exist, are said to behave thus and so does not demonstrate anything about how real people really behave. That an idealized pendulum, which also does not exist, is said to behave thus and so does not demonstrate anything about how actual pendulums behave.

Let us return to the paint company's sample cards. Most people speak of them, and perhaps think of them as samples of paint—the sort of stuff you use to paint the porch. They are not. The cards are infused with inks or dyes of the same color as the paints whose colors they exemplify. It is a fiction that they are samples of paint. But because the sole function of the cards is to convey the colors of the paints, the fiction is no lie. All that is needed is something that is the same colors as the paints. A fiction thus conveys the property we are interested in because in the respect that matters, it is no different from an actual instance. The exemplars need not themselves be paint. Similarly in literary or scientific cases: If the sole objective is to exemplify particular properties, in a suitable context, any symbol that exemplifies those properties will do. If a fiction exemplifies the properties more clearly, simply, or effectively than a strictly factual representation, then it is to be preferred to the factual representation.

Both literary fictions and scientific models exemplify properties and afford epistemic access to them. We discern the properties and can investigate their consequences. Because confounding features have been omitted (the Napoleonic wars in the case of *Pride and Prejudice*, intermolecular attraction in the ideal gas, friction in the model pendulum) we can be confident that the properties we discern in the fictions are due to the factors the fictions make manifest.

Now of course this does not justify a straightforward extrapolation to reality. We cannot reasonably infer from the fact that Elizabeth Bennet was wrong to distrust Mr. Darcy that young women in general are wrong to distrust their suitors. But the grounds for distrust and the reasons those grounds may be misleading are exemplified in the fiction. Once we have seen them clearly there, we may be able to recognize them better in everyday situations. Nor can we reasonably infer from the fact that ideal gas molecules exhibit no mutual attraction, neither do helium molecules. But the behavior exemplified by ideal gas molecules in the model may enable us to recognize such behavior amid the confounding factors that ordinarily obscure what is going on in actual gases.

4. REPRESENTATION-AS

Let us return to Reynolds's representation of Mrs. Siddons as the tragic muse. The tragic muse is a figure from Greek mythology who is supposed to inspire works of tragedy—works that present a sequence of events leading inexorably from a position of eminence to irrecoverable, unmitigated loss, thereby inspiring pity and terror.[10] A tragic muse representation then portrays a figure capable of inspiring such works, one who exemplifies such features as nobility, seriousness, stalwartness, and perhaps a somber dramaticality, as well as a capacity to instill pity and terror. To represent a person as the tragic muse is to represent her in such a way as to reveal or disclose such characteristics in her or to impute such characteristics to her.

An ideal-gas representation is a fiction characterizing a putative gas that would exactly satisfy the ideal gas law. Such a gas is composed of perfectly elastic spherical particles of negligible volume and exhibiting no intermolecular attractive forces. It exemplifies these properties and their consequences, and thereby shows how such a gas would behave. Hughes suggests that the relation between a model and its target is a matter of representation as. The model is a representation—a denoting symbol that has an ostensible subject and portrays its ostensible subject in such a way that certain features are exemplified. It represents its target as exhibiting those features. So to represent helium as an ideal gas is to represent it as composed of molecules having the features exemplified in the ideal gas model—elasticity, mutual indifference, the proportionality of pressure, temperature, and volume captured in the ideal gas law.

Representing a philosophy department as a tree might highlight the way the commitments of the various members branch out of a common, solid, rooted tradition, and the way that the work of the graduate students further branches out from the work of their professors. It might intimate that some branches are flourishing while others are stunted growths. It might even suggest the presence of a certain amount of dead wood. Representing the department as a tree then affords resources for thinking about it, its members and students, and their relation to the discipline in ways that we otherwise might not.

I said earlier that when x represents y as z, x is a z-representation that *as such* denotes y. We are now in a position to cash out the 'as such.' It is because x is a z-representation that x denotes y as it does. x does not merely denote y and happen to be a z-representation. Rather in being a z-representation, x exemplifies certain properties and imputes those properties or related ones to y. The properties exemplified in the z-representation serve as a bridge that connects x to y. This enables x to provide an orientation to its target that affords epistemic access to the properties in question.

Of course, there is no guarantee that the target has the features exemplified by the model, any more than there is any guarantee that a subject represented as the tragic muse has the features exemplified by a painting representing her as the tragic muse. This is a question of fit.

A model may fit its target well or badly or not at all. Like any other case of representation as, the target may have the features exemplified in the model. Then the function of the model is to make those features manifest and display their significance. We may see the target system in a new and fruitful way by focusing on the features that the model draws attention to.

In other cases, the fit is looser. The model may not exactly fit the target. If the features are not the precise features the model exemplifies, they may be relevantly analogous. If gas molecules are roughly spherical and fairly elastic, then we may gain insight into their behavior by representing them as perfectly elastic spheres. Perhaps we will subsequently have to introduce correction factors to accommodate the divergence from the model. Perhaps not. It depends on what degree of precision we want or need. Sometimes, although the target does not quite instantiate the features exemplified in the model, it is not off by much. Where their divergence is negligible, the models, although not strictly true of the phenomena they denote, are true enough of them.[11] This may be because the models are approximately true, or because they diverge from truth in irrelevant respects, or because the range of cases for which they are not true is a range of cases we do not care about, as for example when the model is inaccurate at the limit. Where a model is true enough, we do not go wrong if we think of the phenomena as displaying the features exemplified in the model. Obviously, whether such a representation is true enough is a contextual question. A representation that is true enough for some purposes, or in some respects is not true enough for or in others. This is no surprise. No one doubts that the accuracy of models is limited.

In other cases, of course, the model simply does not fit. In that case, the model affords little or no understanding of its target. Not everyone can be informatively represented as the tragic muse. Nor can every object be informatively represented as a perfectly elastic sphere.

Earlier I dismissed resemblance as a vehicle of representation. I argued that exemplification is required instead. But for *x* to exemplify a property of *y*, *x* must share that property with *y*. So *x* and *y* must be alike in respect of that property. So, it might seem, resemblance in particular respects is what is required to connect a representation with its referent.[12] There is a grain of truth here. If exemplification is the vehicle for representation-as, the representation and its object resemble one another in respect of their exemplified properties. But resemblance, even resemblance in a particular, relevant respect, is not enough, as the following tragic example shows.

On January 28, 1986, the space shuttle *Challenger* exploded because its O-rings failed due to cold weather. The previous day, engineers involved in designing the shuttle had warned NASA about that very danger. They faxed to NASA data to support their concern. The evidence that the O-rings were vulnerable in cold weather was contained in the data. But it was obscured by a melange of other information that was also included.[13] So although the requisite resemblance between the model and target obtained, it was overshadowed in the way that a subtle irregularity in an elaborate tapestry might be. As they were presented, the data instantiated but did not exemplify the vulnerability. They did not represent the O-rings as vulnerable in cold weather. Because the correlation between O-ring degradation and temperature was not perspicuous, the NASA decision makers did not see it. The launch took place, the shuttle exploded, and the astronauts died. When the goal of a representation is to afford understanding, its merely resembling the target in relevant respects is not sufficient. The representation must make the resemblance manifest.

5. PROBLEMS ELUDED

This account evades a number of controversies that have arisen in recent discussions of scientific models. Whether models are concrete or abstract makes no difference. A Tinker Toy model of a protein exemplifies a structure and represents its target as having that structure. An equation exemplifies a relation between temperature and pressure and represents its target as consisting of molecules whose temperatures and pressures are so related. Nor does it matter whether models are verbal or nonverbal. One could represent Mrs. Siddons as the tragic muse in a picture, as Reynolds did, or in a poem as Russell did.[14]

In all cases, models are contrived to exemplify particular features. Theoretical models are designed to realize the laws of a theory.[15] But we should not be too quick to think that they are vacuously true. For by exemplifying features that follow from the realization of the laws, the models may enhance understanding of what the realization of the laws commits the theory to. They may, for example, show that any system that realizes the laws has certain other unsuspected properties as well. The model then can provide reasons to accept or reject the theory. Such a model is a media-tor between the laws and the target system.[16] It in effect puts meat on the bare bones of the theory, makes manifest what its realization requires, and exemplifies properties that are capable of being instantiated in and may be found in the target system. In talking about theoretical models, we should be sensitive to the ambiguity of the word 'of.' Such a model is a model of a theory because it exemplifies the theory. It is a model of the target because it denotes the target. It thus stands in different referential relations to the two systems it mediates between.

Not all models are models of laws or theories. There are phenomenological models as well. These too exemplify features that are ascribed to their target systems. They are streamlined, simplified representations that high-light those properties and exhibit their effects. The difference is that the features phenomenological models exemplify are not captured in laws.

Data models regiment and streamline the data. They impose order on it, by smoothing curves, omitting outliers, grouping together readings that are to count as the same, and discriminating between readings that are to count as different. They thereby bring about the exemplification of patterns and discrepancies that are apt to be obscured in the raw data.

There is evidently no limit on what can be a target. It is commonplace that scientists rarely if ever test theoretical models or phenomenological models against raw data. At best, they test such models against data mod-els. Only data models are apt to be tested against raw data. A theoretical model might take as its target a phenomenological model or a less abstract theoretical model.[17] Then its accuracy would be tested by whether the fea-tures it exemplifies are to be found in the representations that other model provides, and its adequacy would be tested by whether the features found are scientifically significant. We can and should insist that eventually mod-els in empirical sciences answer to empirical facts. But there may be a mul-tiplicity of intervening levels of representation between the model and the facts it answers to.

Because models depend on exemplification, they are selective. A model makes some features of its target manifest by overshadowing or ignoring others. So different models of the same target may make different fea-tures manifest. Where models are thought of as mirrors, this seems to be a problem. It is extremely difficult to see how the nucleus could be mirrored without distortion as a liquid drop and as a shell structure.[18] Because a single material object cannot be both liquid and rigid, there seems to be something wrong with our understanding of the domain if both models are admissible. But if what one model contends is that in some significant respects the nucleus behaves like a liquid drop, and another model con-tends that in some other significant respects it behaves as though it has a shell structure, there is in principle no problem. There is no reason why the same thing should not share some significant properties with liquid drops and other significant properties with rigid shells. It may be surprising that the same thing could have both sets of features, but there is no logical or conceptual difficulty. The models afford different perspectives on the same reality. And it is no surprise that different perspectives reveal different aspects of that reality. There is no perfect model[19] for the same reason that there is no perfect perspective. Every perspective, in revealing some things, inevitably obscures others.

As far as I can see, nothing in this account favors either nominalism or realism. One can run the whole story in terms of properties, as I have done, or in terms of labels. To do it in terms of labels seems perhaps a bit more cumbersome, but even that appears to be a function of familiarity with the devices deployed. Nor does anything in this account favor either scientific realism or antirealism. One can be a realist about theoretical commitments, and take the success of the models to be evidence that there really are such things as charmed quarks. Or one can be an antirealist and take the success of the models to be evidence only of the empirical adequacy of representations that involve charmed-quark-talk. Where we have models that do not exactly fit the data, we can either take an instrumental stance to their function, or take a realist stance and say that the phenomena are a product of signal and noise, and that the models just eliminate the noise. I am not claiming that there are no real problems here, only that the cognitive functions of models that I have focused on do not favor either side of the debates.

6. OBJECTIVITY

A worry remains. The intimate connection that I have sketched between scientific and artistic representations may heighten anxieties about the objectivity of science. I do not think this is a real problem, but I need to say a bit about objectivity to explain why.

We need to distinguish between objectivity and accuracy. A representation is accurate if things are the way it represents them to be. A hunch may be accurate. My completely uninformed guess as to who will win the football game may turn out to be correct. But there is no reason to believe it, for it is entirely subjective. An objective representation may be accurate or inaccurate. Its claim to objectivity turns not on its accuracy, but on its relation to reasons. If a representation is objective, it is assessable by reference to intersubjectively available and evaluable reasons, where a reason is a consideration favoring a contention that other members of the community cannot responsibly reject.[20] Because we are concerned with science here, the relevant community is a scientific community. So scientific objectivity involves answerability to the standards of a scientific community. According to these standards, among the factors that make a scientific result objective are being grounded in evidence, verifiable by further testing, corroborated by other scientists, consistent with other findings, and delivered by methods that have been validated. And generating objective results is what makes a model or method objective.

It is not an accident that my characterization of objectivity is schematic. What counts as evidence, and what counts as being duly answerable to evidence, and who counts as a member of the relevant community are not fixed in the firmament. Answers to such questions are worked out with the growth of a science and the refinement of its methodology. This is not the place to go into the details of such an account of objectivity.[21] What is important for our purposes is this: To be duly answerable to evidence is not necessarily to be directly answerable to evidence. A representation may be abstract so it needs multiple levels of mediating symbols to bring it into contact with the facts. A representation may be indirect. It may involve idealizations, omissions, and/or distortions that have to be acknowledged and accommodated, if we are to understand how it bears on the facts. But if it is objective, then empirical evidence must bear on its acceptability and the appropriate scientific community must be in at least rough accord about what the evidence is (or would be) and how it bears (or would bear) on the representation's acceptability.

I said at the outset that science embodies an understanding of nature. An understanding is a grasp of a comprehensive general body of information that is and manifests that it is responsive to reasons. It is a grasp that is grounded in fact, is duly answerable to evidence, and enables inference, argument, and perhaps action regarding the subject the understanding pertains to. This entails nothing about the way the body of information is encoded or conveyed. Whether symbols are qualitative or quantitative, factual or fictional, direct or oblique, they have the capacity to embody an understanding. To glean an understanding requires knowing how to interpret the symbols that embody it. But we should not think that simply because symbols require interpretation, they are somehow less than objective. So long as there are justifiable, intersubjectively agreed upon standards of interpretation, objectivity is not undermined. So although scientific models do not accurately mirror anything in nature, they are capable of affording understanding of what occurs in nature.[22]

NOTES

1. This use of 'denote' is slightly tendentious, first because denotation is usually restricted to language, second because even within language it is usually distinguished from predication. As I use the term, predicates and generic nonverbal representations denote the members of their extensions. See Elgin (1983, pp. 19–35).
2. B. Russell (1968, p. 41).
3. Goodman (1968, pp. 21–26).
4. Suárez (2004b).
5. Hughes (1997).
6. Goodman (1968, p. 4); Suarez (2003).

7. Goodman (1968, pp. 45–68); Elgin (1996, pp. 171–183).
8. 'Aristotelian Natures and Modern Experimental Method' in Cartwright (1999a, pp. 77–104).
9. Jane Austen, Letter to her niece, Anna Austen Lefroy, September 9, 1814, in *Letters of Jane Austen*, Bradbourn Edition, www.pemberley.com/janeinfo/brablets.html. Consulted May 4, 2005.
10. Aristotle, *Poetics*, Book 6, lines 20–30. In McKeon (1973, p. 677).
11. See Elgin (2004).
12. This is the position Giere takes about the relation between a model and its target system. See Giere (1999).
13. Tufte (1997, pp. 17–31).
14. W. Russell, 'The Tragic Muse: A Poem Addressed to Mrs. Siddons,' 1783. www.dulwichpicturegallery.org.uk/collection/search/display.aspx?im=252. Consulted January 12, 2006.
15. Giere (1999, p. 92).
16. Morrison and Morgan (1999).
17. Suárez (2003, p. 237).
18. This example comes from Frigg and Hartmann (2006, p. 3).
19. Teller (2001).
20. See Scanlon (1998, pp. 72–75). I say 'assessible by reference to reasons' rather than 'supportable by reasons' because an objective judgment may not stand up. If I put forth my judgment as an objective judgment, submit it to a jury of my peers, it is objective, even if my peers repudiate it.
21. For the start of such an account see Scheffler (1982).
22. I would like to thank Israel Scheffler, Nancy Nersessian, John Hughes, and the participants in the 2006 Workshop on Scientific Representation at the Universidad Complutense de Madrid for helpful comments on earlier drafts of this chapter.

6 Explanatory Fictions

Alisa Bokulich

> They [semiclassical theorists] do not describe an electron as a point particle moving in real time along a classical orbit. Nevertheless, the fiction can be productive, permitting one to talk about classical orbits as if they were real. (Kleppner & Delos, 2001, p. 606)

1. INTRODUCTION

The preceding epigraph sounds like it could have been taken from a page of Hans Vaihinger's (1911/1952) book on fictionalism, *The Philosophy of 'As If.'*[1] Instead, however, it comes from a recent article in a physics journal, coauthored by an experimental physicist at MIT (Dan Kleppner) and a theoretical physicist at the College of William and Mary (John Delos). Like Vaihinger, these physicists are defending the view that some fictions have a legitimate role to play in science, and also like Vaihinger they are defending these fictions on pragmatic grounds. Kleppner and Delos are concerned specifically with an area of research known as semiclassical mechanics, and the remarkable fertility of using fictional classical electron orbits to describe the quantum spectra of atoms placed in strong external fields. A striking feature of this research is that these fictional orbits are not simply functioning as calculational devices, but also seem to be playing a central role in the received scientific explanation of these phenomena. Although there is a growing recognition that fictions have some legitimate role to play in scientific theorizing, one function that is traditionally denied to fictions is that they can *explain*. Even Vaihinger, who argues for the pervasiveness of fictions in science, rejects the view that there are explanatory fictions.

In what follows, I defend the view that, in some cases, fictions can genuinely explain. I begin, in section 2, by situating my approach to fictions in the context of Vaihinger's classic account, then, in section 3, turn to the concrete case of fictional classical orbits explaining quantum spectra in modern semiclassical mechanics. In section 4, I introduce a new philosophical account of scientific explanation that I argue is capable of making sense of the explanatory power of fictions. Finally, I conclude in section 5 by relating this case study and the new philosophical account of scientific explanation back to Vaihinger's account of fictions.

2. BEYOND VAIHINGER'S ACCOUNT OF FICTIONS

In his comprehensive account of fictions in human thought, Vaihinger identifies four key features that he takes to characterize fictions.[2] The first and most salient characteristic of fictions is their deviation from reality. That is, fictions involve what he calls a "contradiction" with reality. He introduces an elaborate taxonomy of different kinds of fictions based on the various ways in which this deviation from reality can take place (e.g., neglective fictions, heuristic fictions, and abstract generalizations) as well as based on the various subject matters in which fictions can be employed (e.g., juristic fictions, ethical fictions, mathematical fictions and scientific fictions). In defining a fiction as almost any deviation from reality, however, Vaihinger's account does not clearly distinguish among discarded theories, models, idealizations, abstractions, and what we might today more narrowly call fictions. Although there may be no hard and fast distinctions between these various things that Vaihinger groups together as 'fictions,' when it comes to assessing their epistemic status, such distinctions might nonetheless turn out to be important. So, for example, although phlogiston and frictionless planes are both "fictions" on Vaihinger's account, we might want to say that the latter is an idealization that can be subject to something like what Ernan McMullin (1985) calls a de-idealization analysis. Hence, there is an important sense in which reasoning based on frictionless planes has a different epistemic status and involves a different sort of justification than reasoning on the basis of a genuinely fictional entity of a discarded theory, such as in the case of phlogiston.

The second essential characteristic of fictions that Vaihinger identifies is that they are ultimately to be eliminated. He writes, "the fiction is a mere auxiliary construct, a circuitous approach, a scaffolding afterwards to be demolished" (Vaihinger, 1911/1952, p. 88). Elsewhere he notes that a fiction "only falsifies reality with the object of discovering the truth" (Vaihinger, 1911/1952, p. 80). That is, the proper aim and function of fictions is to prepare the road to truth—not be a permanent stand-in for truth. One might have assumed that a recognition of the pervasiveness of fictions in science would lead Vaihinger to embrace some form of antirealism, such as instrumentalism. This second condition that he places on fictions, however, reveals that this is not the case; Vaihinger's view is in many respects closer to a form of scientific realism. Indeed, he is quite optimistic that as our experience becomes richer and our scientific methods refined, these various fictions can and will be eliminated.

There are two important questions to separate here. The first question is whether fictions are eliminable from scientific practice. One might reasonably argue that all scientific representation involves some sort of idealization, abstraction, or fictionalization of the target system. This is not to say, of course, that scientists cannot recognize some scientific models as being more or less idealized or fictionalized than others. Hence, pace

Vaihinger, fictions are not ultimately to be eliminated, but rather they are a permanent feature of science. There is, however, another possible interpretation of Vaihinger here, and that is that he did not mean that all fictions are eliminable from science, but only that any *given* fiction will someday be eliminated. Although I think this view is more defensible, it is not clear that even this more restricted claim will always be the case. As I shall suggest later, there may be some fictions in science that, because of their great utility and fertility, scientists may decide to always keep "on the books," even though there may be less fictionalized hypotheses available for that same phenomenon.

Even if we grant that fictions and idealizations are a pervasive and permanent feature of science, it is a second, distinct question, whether such an ineliminability of fictions necessarily undermines scientific realism. In other words, just because a scientific theory or model involves a fictionalization, does that mean that it cannot give us genuine insight into the way the world is? Although I cannot defend these theses in generality here, my own view is that the answer to both of these questions is no: All scientific representation *does* involve some idealization or fictionalization, and this does not in and of itself render scientific realism untenable.

The third key characteristic of fictions that Vaihinger identifies is that there should be an "express awareness that the fiction is just a fiction" (p. 98). That is, when scientists deploy a fiction, they do so knowing full well that it is a false representation. I think this is actually a very import feature of scientific fictions. Much of our discomfort with the idea that there are fictions in science stems from our concern that fictions will necessarily lead scientists astray and render science subjective and arbitrary. As I think Vaihinger rightly recognized, however, this need not be the case. As long as scientists deploy a fiction with the full knowledge that it is just a fiction, then it is much less likely that they will be misled by it. Furthermore, the recognition that some concept is a fiction need not make science subjective at all; indeed, the fiction being deployed can be objectively recognized as a fiction by the scientific community as a whole. Finally, the use of fictions also need not render science arbitrary. As Vaihinger repeatedly emphasizes, scientific fictions are constrained by their utility and expediency.

Vaihinger makes a very helpful distinction between a fiction and an hypothesis. With a hypothesis, the scientist is not yet sure whether it is an accurate representation of the object, system or process of interest. As Vaihinger explains, "An hypothesis is directed toward reality, i.e. the ideational construct contained in it claims, or hopes, to coincide with some perception in the future. It submits its reality to the test and demands *verification*" (p. 85; emphasis original). By contrast, fictions make no claim to truth; rather than being subject to verification, Vaihinger argues they should only be subject to *justification*.[3] That is, fictions are to be judged by their utility and expediency.

This emphasis on the pragmatic function of fictions becomes Vaihinger's fourth key characteristic, which separates out those fictions that deserve the label "scientific" from those that do not. He writes,

> Where there is no expediency the fiction is unscientific. . . . [Hume's] idea of the 'fiction of thought' was that of a merely subjective fancy, while ours . . . includes the idea of its utility. This is really the kernel of our position, which distinguishes it fundamentally from previous views. (p. 99)

Fictions have a legitimate role to play in science insofar as they are useful in furthering the aims and goals of science. Although Vaihinger does not spell this out in any great detail, his approach does suggest that there may be a variety of ways in which fictions can be pragmatically useful. For example, some fictions may be useful as proto-theories, other fictions useful as calculational devices, and still other fictions useful in generating predictions.

One function that Vaihinger clearly denies to fictions, however, is that they can explain. Drawing on his distinction between hypotheses and fictions Vaihinger writes, "The hypothesis results in real explanation, the fiction induces only an illusion of understanding" (p. xv). The reason, he explains, is that "[E]very fiction has, strictly speaking, only a practical object in science, for it does not create real knowledge" (p. 88). In other words, explanation and understanding are not to be counted among the ends of science for which fictions can be expedient, precisely because explanation requires having genuine insight into the way the world is, and fictions are incapable of giving us this sort of insight. So, for example, although the Descartes vortex model of the solar system might make us feel like we have understood why all the planets move in the same direction around the sun, this understanding is illusory, and no genuine explanation of planets' motion has been given.

The view that Vaihinger is expressing here regarding the explanatory impotence of fictions is a widespread and intuitively plausible one. Nonetheless, I think it is mistaken. Although it is certainly not the case that all fictions can explain, I believe that some fictions can give us genuine insight into the way the world is, and hence be genuinely explanatory and yield real understanding. I shall call this (proper) subset of fictions *explanatory fictions*, and distinguish it from what we might call *mere* fictions. In the next section I shall show that it is just such an explanatory fiction that Kleppner and Delos (2001) are calling attention to in the quotation given as the epigraph of this chapter.

3. THE CASE OF CLASSICAL ORBITS AND QUANTUM SPECTRA

Although it was known from the early 19th century that different elements (such as hydrogen, helium, sodium) can absorb and emit light only at a specific set of frequencies, yielding a "signature" line spectra, the first successful explanation of these spectral lines for the simplest element, hydrogen, did not occur until Niels Bohr introduced his planetary model of the atom in a trilogy of papers published in 1913. Bohr proposed that the atom consists of a dense nucleus, where most of the mass of the atom is concentrated, and the electrons orbit this nucleus in a discrete series of allowed concentric rings, known as stationary states. The state is "stationary" in the sense that when the electron is traveling along one of these orbits its energy does not change. Instead, the atom can only gain (or lose) energy when the electron jumps from one allowed orbit to another. Bohr was able to show for the hydrogen atom that each spectral line corresponds to a particular jump of the electron from one allowed orbit to another. Moreover, in his famous "correspondence principle" Bohr proposed that *which* quantum jumps between stationary states were allowed was determined by the nature of the classical motion of the electron along the relevant orbital trajectory.[4]

When Bohr first proposed his model of the atom, he introduced the idea that electrons in atoms follow classical trajectories as a *hypothesis* in Vaihinger's sense. In other words, he took this to be potentially a literal description of the behavior of electrons in an atom. Despite the remarkable successes of Bohr's model in explaining the hydrogen spectrum as well as many other quantum phenomena, by the end of that decade there was mounting empirical evidence that this hypothesis that electrons in atoms are following definite trajectories was problematic. Not only did Bohr's old quantum theory have difficulty in calculating the energies of more complicated elements such as helium, but it also seemed unable to account for the fact that when atoms are placed in a magnetic field, the individual spectral lines are split into a complex multiplet of lines known as the "anomalous Zeeman effect." Right before the overthrow of Bohr's old quantum theory, and its replacement by modern quantum mechanics, Wolfgang Pauli wrote,

> How deep the failure of known theoretical principles is, appears most clearly in the multiplet structure of spectra. . . . One cannot do justice to the simplicity of these regularities within the framework of the usual principles of the [old] quantum theory. It even seems that one must renounce the practice of attributing to the electrons in the stationary states trajectories that are uniquely defined in the sense of ordinary kinematics. (Pauli, 1925/1926, p. 167; quoted in Darrigol, 1992, pp. 181–182)

However, even before the rejection and replacement of Bohr's model of the atom by the new quantum theory in 1925, the idea that electrons are following definite trajectories in atoms had already begun to be transformed by the scientific community from a hypothesis to a useful *fiction*.

For example, in a 1920 article published in *Nature*, Norman Campbell wrote of Bohr's model of the atom,

> Nor is it [the assumption that electrons are *not* moving] physically impossible if we accept Bohr's principle of 'correspondence,' which has been so astoundingly successful in explaining the Stark effect [splitting of spectral lines in an electric field] and in predicting the number of components in lines of the hydrogen and helium spectra. According to that principle, the intensity and polarisation of components can be predicted by the application of classical dynamics to certain assumed orbits, although it must be assumed at the same time that the electrons are *not* moving in those orbits. If intensity and polarisation can be predicted from orbits that are wholly fictitious, why not energy? (Campbell, 1920, p. 408)

Campbell here is defending the practical utility of the idea of electrons moving in classical trajectories in atoms, even though he believes that these trajectories must be regarded as "wholly fictitious."

Even Bohr, as early as 1919, seemed no longer to view the motion of electrons in concentric stationary states as a literal description. Indeed, he wrote to a colleague that year, "I am quite prepared, or rather more than prepared, to give up all ideas of electronic arrangements in rings" (Bohr to O. W. Richardson, 25 December 1919; quoted in Heilbron 1967, p. 478). It is interesting, however, that he rejected Campbell's label of "wholly fictitious" as a correct description of the status of these classical electron orbits. In a reply to Campbell also published in *Nature* Bohr wrote,

> I naturally agree that the principle of correspondence, like all other notions of the [old] quantum theory, is of a somewhat formal character. But, on the other hand, the fact that it has been possible to establish an intimate connection between the spectrum emitted by an atomic system, deduced . . . on the assumption of a certain type of motion of the particles in the atom . . . appears to me to afford an argument in favour of the reality of the assumptions of the spectral theory of a kind scarcely compatible with Dr. Campbell's suggestions. (Bohr, 1921, pp. 1–2)

In other words, although Bohr takes the classical electron orbits to be only a "formal description," he does think that they nonetheless give real insight into the structure and behavior of atoms, and hence are not properly thought of as *wholly* fictional.

With the advent of the new quantum theory, the idea that electrons are actually following definite classical trajectories in atoms would be entirely eliminated. Indeed, Heisenberg's uncertainty principle, introduced in 1927, would show that quantum particles, such as electrons, cannot have a precise position and a precise momentum at the same time, as would be required by the classical notion of a trajectory. Rather than having either a static or moving position, the electron is now more properly thought of as a cloud of probability density around the nucleus of the atom. Surprisingly, however, the introduction of modern quantum mechanics did not in fact mark the end of this history of describing electrons in atoms as following definite classical trajectories. Ironically, it was the discovery of a new generation of "anomalous" spectral data in the late 1960s that would lead to the reintroduction of the notion of classical electron trajectories in atoms—though this time with the express recognition that these classical trajectories were nothing more than useful fictions.

Although the behavior of ordinary atoms in relatively weak external magnetic and electric fields is well understood, when one examines the behavior of highly excited atoms (known as Rydberg atoms), in very strong external fields surprising new phenomena occur.[5] In a series of experiments beginning in 1969, William Garton and Frank Tomkins at the Argonne National Laboratory examined the spectra of highly excited barium atoms in a strong magnetic field. When the magnetic field was off, these Rydberg atoms behaved as expected: As the energy of the photons being used to excite the atom increased, there was a series of peaks at the energies which the barium atom could absorb the photons; when the ionization energy was reached (that is, the energy at which the outer electron is torn off leaving a positive ion), there were no more peaks in the absorption spectrum, corresponding to the fact that the barium atom could no longer absorb any photons. However, when the authors applied a strong magnetic field to these barium atoms and repeated this procedure, a surprising phenomenon occurred: The barium atoms continued to yield absorption peaks long after the ionization energy had been reached and passed (see Figure 6.1).

These oscillations in the spectrum were later named "quasi-Landau" resonances, and were shown to have a spacing independent of the particular type of atom. Remarkably, even almost 20 years after these quasi-Landau resonances were first discovered, a full theoretical explanation of them remained an outstanding problem.

The situation was further exacerbated by the fact that experimentalists were continuing to find new resonances above the ionization limit. For example, higher resolution experiments on a hydrogen atom in a strong magnetic field, performed by Karl Welge's group in Bielefeld in the mid-1980s, revealed many more types of resonances in the absorption spectrum (Main et al., 1986; Holle et al., 1988).[6] Furthermore, these new resonances seemed to have lost the regularity of the quasi-Landau resonances discovered earlier. Instead, these new high-resolution spectral data exhibited a complex irregular pattern of lines. By the end of the 1980s, Kleppner and his colleagues wrote,

"A Rydberg atom in a strong magnetic field challenges quantum mechanics because it is one of the simplest experimentally realizable

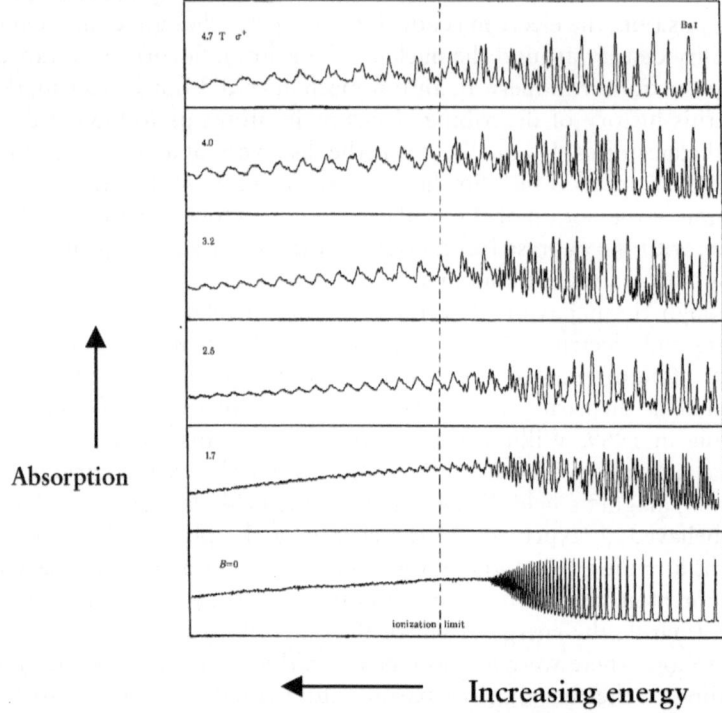

Figure 6.1 The absorption spectrum of barium. Higher energies are to the left, and the vertical dashed line is the ionization threshold, energies above which the barium atom ionizes. The bottom row is the spectrum with no magnetic field (*B* = 0), and the subsequent rows above that are the spectra with a magnetic field at 1.7, 2.5, 3.2, 4.0, and 4.7 Tesla, respectively. Note the surprising oscillations in the absorption spectrum above the ionization limit when a strong magnetic field is present. (Adapted from Lu et al., 1978, Figure 1, with permission.)

systems for which there is no general solution. . . . We believe that an explanation of these long-lived resonances poses a critical test for atomic theory and must be part of any comprehensive explanation of the connection between quantum mechanics and classical chaos." (Welch et al., 1989, p. 1975)

Once again the Zeeman effect was yielding anomalous spectra, whose explanation seemed to require the development of a new theoretical framework. Although modern quantum mechanics was never in doubt, the mesoscopic nature of Rydberg atoms suggested that an adequate theoretical explanation of these resonance phenomena would require not only quantum mechanics, but concepts from classical chaos as well.

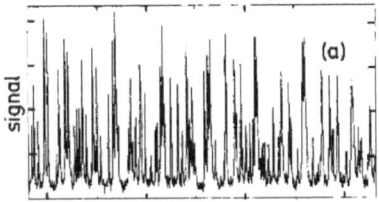 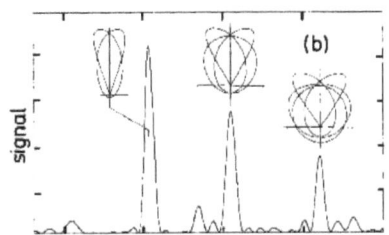

Figure 6.2 (a) The irregular-looking scaled energy spectrum for a hydrogen atom near the ionization limit in a strong magnetic field. (b) The Fourier transform of this same spectrum into the time domain. The particular classical closed orbit corresponding to each of these well-defined peaks is superimposed. (From Holle et al. [1988, Figure 1]; courtesy of J. Main. Copyright 1988 by the American Physical Society.)

An important step toward explaining these resonances was made by the Bielefeld group in a subsequent paper. They realized that if one takes the Fourier transform of the complex and irregular looking spectra, an orderly set of strong peaks emerges in the time domain.

This resulting "recurrence spectrum" revealed that the positions of these peaks in the time domain were precisely at the periods, or transit times, of the classically allowed closed orbits for the electron moving in the combined Coulomb and magnetic fields; that is, each peak in the quantum spectrum corresponds to a different closed classical trajectory. The Bielefeld group wrote,

> Though those experiments [by Welge's group in the 1980s] suggested the existence of even more resonances their structure and significance remained fully obscure. In this work we have discovered the resonances to form a series of strikingly simple and regular organization, not previously anticipated or predicted. . . . *The regular type resonances can be physically rationalized and explained by classical periodic orbits of the electron on closed trajectories* starting at and returning to the proton as origin with an orbital recurrence-time T characteristic for each v-type resonance. (Main et al., 1986, pp. 2789–2790; emphasis added)[7]

Note that the explanation being offered for these anomalous resonances and their regular organization makes explicit appeal to the fictional assumption that these Rydberg electrons, instead of behaving quantum mechanically, are following definite classical trajectories.

The appeal to fictional classical trajectories in explaining these quantum spectra is not an isolated misstatement by Karl Welge and his colleagues; rather, it became the foundation for the received scientific explanation developed two years later by Delos and his student Meng-Li Du (Delos & Du, 1988; Du & Delos, 1988). The correspondence between each of the positions of these peaks in the spectrum and the transit times of the electron on a particular classical closed trajectory suggested that a semiclassical approach such as Gutzwiller's periodic orbit theory would be the key to explaining the experimental data (Gutzwiller, 1971; 1990). Their theory, which is known as "closed orbit theory," is similar to Gutzwiller's in that it is grounded in a semiclassical approximation to the Green's function, though instead of using periodic orbits, Delos's theory makes use of closed orbits, namely, those orbits that are launched from, and return to, the vicinity of the nucleus.

In order to use closed orbit theory, one must first use classical mechanics to calculate the allowed orbits of a charged classical particle moving under the action of the combined Coulomb and magnetic field. These closed orbits can exhibit a variety of loops and zigzags before returning to the nucleus. It turns out that, of all the possible allowed closed orbits of an electron in such a field, only about 65 orbits are relevant to explaining the quantum spectrum (Du & Delos, 1988, p. 1906). Which orbits are relevant, and how they explain the anomalous spectra as Du and Delos claim in the material just described, has been summarized in an intuitive way as follows:

> Delos's insight was to realize that interpreting the departing and arriving electron as a wave meant that its outgoing and incoming portions will inevitably display the symptoms of interference. . . . [T]he survival of some of these quantum mechanical waves and the canceling out of others result in only certain trajectories' being allowed for the electron. . . . Once Delos established that only some trajectories are produced, he had effectively explained the new mechanism that caused the mysterious ripples [in the absorption spectrum above the ionization limit]. The Rydberg electron is allowed to continue to absorb energy, so long as that energy is precisely of an amount that will propel the electron to the next trajectory allowed by the interference pattern. (von Baeyer, 1995, p. 108)

It is worth emphasizing again that this explanation of the anomalous resonances in the spectra is not a purely quantum explanation, deducing the spectrum directly from the Schrödinger equation.[8] Rather, it involves a careful blending of quantum and classical ideas: On the one hand the Rydberg electron is thought of quantum mechanically as a wave exhibiting the phenomenon of interference, while also being thought of fictionally as a particle following specific classical closed orbit trajectories.

Despite the unorthodox hybridization of classical and quantum ideas in this explanation, closed orbit theory has proven to be strikingly successful empirically. With these classical closed orbits, one can predict the wavelength, amplitude, and phase of these resonances to within a few percent, and, furthermore, the predictions of this theory have proven to be in very close agreement with the data generated by numerous subsequent experiments on the absorption spectra of hydrogen, helium, and lithium atoms in strong magnetic fields.[9] This striking success of closed orbit theory shows that even in atomic physics, which is clearly under the purview of quantum mechanics, it is classical mechanics as developed through modern semiclassics that is proving to be the appropriate theoretical framework for explaining many of these quantum phenomena.

Closed orbit theory not only can explain the particular details of the experimental spectra, but can also explain why the earlier, lower-resolution data of Tomkins and Garton yielded a very orderly series of oscillations, whereas the later, higher-resolution data of Welge and colleagues revealed a wildly irregular series of oscillations. The explanation, once again, rests on a thorough mixing of classical and quantum ideas—specifically, a mixing of the quantum uncertainty principle with the fact that classical chaos is a long-time ($t \rightarrow \infty$) phenomenon that on short time scales can still look orderly. Because the low-resolution experiments involved only a rough determination of the energy, only the short-time classical dynamics is relevant to the spectrum. The high-resolution experiments, by contrast, involved a more precise determination of energy, and hence the longer time dynamics of the classical system is relevant. Because the long-time dynamics of a classical electron in a strong magnetic field is chaotic, this complexity manifests itself in the spectra.[10]

Ten years after closed orbit theory was introduced, Delos, Kleppner, and colleagues showed that not only can classical mechanics be used to generate the quantum spectrum, but, even more surprisingly, the experimental quantum spectrum can be used to reconstruct the classical trajectories of the electron. As emphasized earlier, part of the reason this is surprising is that electrons do not, in fact, follow classical trajectories at all—they are fictions. Recognizing this tension, they write, "We present here the results of a new study in which semiclassical methods are used to reconstruct a trajectory from experimental spectroscopic data. When we speak of the 'classical trajectory of an electron,' we mean, of course, the path the electron would follow if it obeyed the laws of classical mechanics" (Haggerty et al., 1998, p. 1592). Although the previous experiments could be used to establish the actions, stabilities, and periods of the closed orbits, they could not be used to determine the orbits themselves, that is, the electron positions as a function of time. In this paper, however the authors show how "by doing spectroscopy in an oscillating field, we gain new information that allows us to reconstruct a trajectory directly—without measuring the wave function and without relying on detailed knowledge of the static Hamiltonian" (Haggerty et al., 1998, p. 1592).

Their experiment involves examining the spectrum of a highly excited (that is, Rydberg) lithium atom in an electric field (a phenomenon known as the Stark effect). Although the behavior of a hydrogen atom in an electric field is regular, the behavior of a lithium atom in an electric field can be chaotic. Using an extension of closed orbit theory, the investigators were able to show that an oscillating electric field reduces the strength of the recurrences in the spectrum, that is, the heights of the peaks, in a manner that depends on the Fourier transform of the classical electron orbits in the static electric field. Hence, by experimentally measuring the Fourier transform of the motion for a range of frequencies, one can then take the inverse Fourier transform, and obtain information about the electron's orbits. Using this technique, they were able to successfully reconstruct, from the experimental *quantum* spectra, two *classical* closed orbits of an electron in an electric field (these orbits are referred to as the "2/3" and "3/4" orbits, and are pictured in Figure 6.3).

They concluded, "Our experiment produces accurate, albeit low-resolution, pictures of classical trajectories important to the Stark spectrum of lithium" (Haggerty et al., 1998, p. 1595). Although they used the Stark spectrum of Rydberg lithium in this experiment, their method of extracting classical trajectories from quantum spectra can be applied to a variety of other systems. They emphasize that the limits they encountered in resolving these trajectories are experimental, not fundamental, being many orders of magnitude away from the limits imposed by Heisenberg's uncertainty principle. Hence, although these experiments are deriving pictures of classical trajectories from quantum spectra, these trajectories in no way undermine the uncertainty principle.

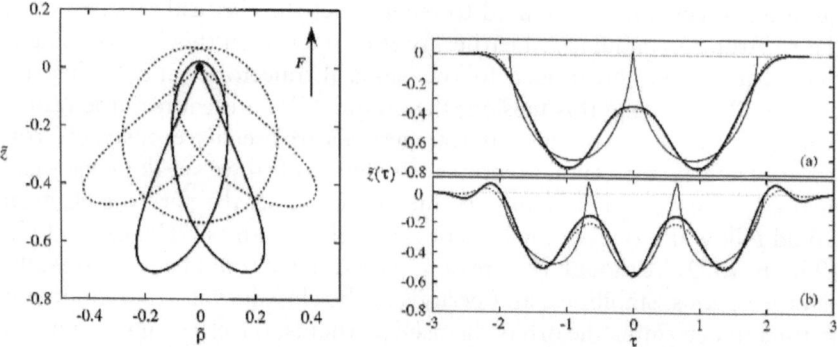

Figure 6.3 On the left: Two classical closed orbits of a Rydberg electron in an electric field; the solid line is the "2/3" orbit and the dotted line is the "3/4" orbit. The dark dot represents the nucleus of the atom. On the right: (a) the "2/3" orbit and (b) the "3/4" orbit. The light solid lines show the exact classical trajectories and the heavy lines are the experimentally reconstructed trajectories. Because the experimental frequency range is limited, the exact trajectories filtered through the experimental frequency window have also been included as the dashed lines for comparison. (From Haggerty et al., 1998, Figure 2 and Figure 4; courtesy of D. Kleppner. Copyright 1998 by the American Physical Society.)

Underlying these experiments is the following pressing but unspoken question: Given that classical mechanics is *not* true, and that electrons in atoms do *not* actually follow definite trajectories, how can one legitimately speak of experimentally measuring such trajectories from a quantum spectrum at all? In a more recent paper, Kleppner and Delos (2001) tackle head on this question of the reality of these trajectories. They write,

> Because of the power of the concept of periodic orbits, one might question as to what extent they 'really exist' in the atom. Insight into this question was provided by a recent experiment at MIT, in which recurrence spectroscopy was used to measure the time dependence of an electron's motion along one of the closed orbits. To put it more precisely, recurrence spectroscopy was used to measure the time dependence of the fictitious classical trajectory that can be used as a calculational device to construct the quantum propagator. And to put it less precisely, recurrence spectroscopy showed how an electron in an atomic system would move in space and time if it obeyed classical physics. (Kleppner & Delos, 2001, p. 606)

After discussing their experiment in more detail, however, Kleppner and Delos seem tempted by the view that these electron trajectories are more than mere fictions or calculational devices: They write, "These results lead us to question whether a trajectory should be described as truly 'fictitious' if one can measure its detailed properties" (Kleppner & Delos, 2001, p. 610). The full realist claim, that electrons in atoms *really are* following these definite classical trajectories, would amount to a rejection of modern quantum mechanics and a violation of Heisenberg's uncertainty principle; this is not something that semiclassical theorists, Kleppner and Delos included, intend to do.[11]

Nonetheless, I think Kleppner and Delos are groping toward an important point here: Not all fictions are on par with each other. Even if we bracket those fictions that are useless for science and restrict our attention to those fictions that, as Vaihinger says, justify themselves by their utility and expediency, there are still some important distinctions to be made. On the one hand there are fictions in the sense of what Kleppner and Delos call "mere calculational tools." An example of such "mere fictions" might be Ptolemaic astronomy with its epicycles. Although Ptolemaic astronomy might be a useful calculational tool for some navigators and surveyors, no one thinks that these fictions are giving any real insight into the way the world is or offering any explanations. On the other hand, however, there are some fictions in science that go beyond being simply calculational devices. These fictions, by contrast, do give some genuine insight into the way the world is and do seem to have some genuine explanatory power. I have called these latter sort of fictions *explanatory fictions*. As an example of an explanatory fiction, Kleppner and Delos cite rays of light. They write, "When one sees the sharp shadows of buildings in a city, it seems difficult to insist that light-rays are merely calculational tools that provide approximations to the full solution of the wave equation" (Kleppner & Delos, 2001, p. 610). Although they can certainly be used as calculational tools, these latter sort of fictions also carry explanatory force, and correctly capture in their fictional representation real features of the phenomena under investigation.

4. HOW FICTIONS CAN EXPLAIN

The chief obstacle to admitting the existence of explanatory fictions is that it is difficult to imagine how a fiction could possibly explain. Indeed, on the two most widely received philosophical accounts of scientific explanation— Carl Hempel's deductive-nomological (D-N) account and Wesley Salmon's causal-mechanical account—fictions cannot explain at all. According to Hempel (1965), a scientific explanation is essentially a deductive argument, where the phenomenon to be explained—the "explanandum"—is shown to be the deductive consequence of premises describing the relevant law or laws of nature and any relevant initial conditions (these premises that do the explaining are collectively referred to as the "explanans"). In order to count as a genuine scientific explanation, one of the further conditions that Hempel imposes is what he calls the "empirical condition of adequacy," by which he means specifically that the sentences constituting the explanans must be entirely true (Hempel, 1965, p. 248). This is not "empirical adequacy" in our modern parlance, but rather a condition of *Truth*—with a capital "T." Hempel makes it quite clear that it is insufficient for the explanans to be merely "highly confirmed by all of the relevant evidence available" (Hempel, 1965, p. 248). Given this strict requirement of truth, it is clear that fictions cannot be explanatory on this account of scientific explanation.

The second most widely received account of scientific explanation, Salmon's causal-mechanical account, also rules out the possibility that fictions can explain. On Salmon's (1984) account, to explain a phenomenon or event is to describe the causal-mechanical processes that led up to, or constitute, that phenomenon or event. Salmon draws a sharp distinction between genuine causal processes, which are physical processes capable of "transmitting a mark," and what he calls "psuedo-processes," which cannot. Only the former can lead to genuine scientific explanations. Thus in Salmon's account, like Hempel's, it seems that fictions cannot genuinely explain. Fictional entities and fictional processes do not meet the requirements of a genuine physical processes capable of transmitting a mark. Put more simply, a fiction A cannot be the cause of some phenomenon B—and hence explain B—if A does not exist.

The question, then, is whether we should dismiss the explanation being offered by Delos and colleagues of the resonances in the quantum spectra in term of fictional closed orbits as no explanation at all because it does not fit our preconceived philosophical ideas about scientific explanation. The answer, I believe, is no. Although the closed orbit explanation of the spectra is not entirely true, as required by Hempel's account, nor can the fictional orbits be properly thought of as the *cause* of the oscillations, as required by Salmon's account, we should take the actual explanatory practices of scientists seriously, and nonetheless recognize it as a distinctive form of scientific explanation. Elsewhere (Bokulich, 2008a) I have developed an alternative account of scientific explanation, called "model explanations," that not only can make sense of the explanatory power of idealized scientific models, but, as I shall argue next, also can correctly describe the sort of explanation that is being offered in the present case by closed orbit theory.

Model explanations can be characterized by the following three core features. First, the explanans must make essential reference to a scientific model, and that model (as I believe is the case with all models) involves some idealization and/or fictionalization of the system it represents. Second, that model is taken to explain the explanandum by showing that the pattern of counterfactual dependence in the model mirrors in the relevant respects the pattern of counterfactual dependence in the target system. Following James Woodward (2003), this pattern of counterfactual dependence can be explicated in terms of what he calls "what-if-things-had-been-different questions," or w-questions for short. That is, "the explanation must enable us to see what sort of difference it would have made for the explanandum if the factors cited in the explanans had been different in various possible ways" (Woodward, 2003, p. 11). Although I think that Woodward's approach is largely right, where I part company with his view is in his construal of this counterfactual dependence along strictly manipulationist or interventionist lines. It is this manipulationist construal that restricts Woodward's account to purely *causal* explanations, and, as I argued earlier, I think it is a mistake to construe all scientific explanation as a species of causal explanation. The third condition that a model explanation must satisfy is that there must be what I call a further justificatory step. Very broadly, we can understand this justificatory step as specifying what the domain of applicability of the model is, and showing that the phenomenon in the real world to be explained falls within that domain. Although the details of this justificatory step will depend on the details of the particular model in question, it typically proceeds either from the ground up, via something like a de-idealization analysis of the model (McMullin, 1985), or top down via an overarching theory that justifies the modeling of that domain of phenomena by that idealized model.

It turns out that there are a variety of different subspecies of model explanations (Bokulich, 2008a); hence even after one has identified a particular scientific explanation as a model explanation, there still remains the question of what *type* of model explanation it is. Determining the type of model explanation requires articulating what might be called the source of this counterfactual dependence. The type of model explanation that I believe is most relevant to closed orbit theory (and semiclassical explanations more generally—see also Bokulich, 2008b) is what I have called *structural model explanations*. Following Peter Railton (1980, Section II.7) and R. I. G. Hughes (1989), a *structural explanation* is one in which the explanandum is explained by showing how the (typically mathematical) structure of the theory itself limits what sorts of objects, properties, states, or behaviors are admissible within the framework of that theory, and then showing that the explanandum is in fact a consequence of that structure.[12] A *structural model explanation*, then, is one in which not only does the explanandum exhibit a pattern of counterfactual dependence on the elements represented in the model, but, in addition, this dependence is a consequence of the structural features of the theory (or theories) employed in the model.

Applying this framework to the example presented in the previous section, I argue that classical closed orbits—despite being fictions—are able to genuinely explain the oscillations in the absorption spectrum in the sense that they provide a structural model explanation of this phenomenon. First, the closed classical orbits are the fictional elements that make up the semiclassical model of the quantum spectra. Second, there is a pattern of counterfactual dependence of the oscillations in the spectrum on the various features of these closed orbits (i.e., their actions, stabilities, periods, etc.). Moreover, this counterfactual dependence allows one to correctly answer a wide range of what-if-things-had-been-different questions. For example, one can say exactly how the oscillations peaks would have been different if the closed trajectories had been altered in various sorts of ways. Third, there is a "top-down" justificatory step provided by closed orbit theory, which specifies precisely how these classical trajectories can be legitimately used to model the quantum phenomena.[13] The justification provided by closed orbit theory and the wide range of w-questions that closed orbit theory can correctly answer together suggest that these classical trajectories—despite their fictional status—are nonetheless giving us real insight into the structure of the quantum phenomena. In other words, although these classical trajectories are also useful calculational tools, they are not *mere fictions*. Insofar as these closed orbits are giving us genuine insight into the structure of the quantum dynamics, they are *explanatory fictions*.

5. CONCLUSION

It is instructive to examine to what extent the present example of fictional classical orbits in quantum spectra fits with Vaihinger's account, and to what extent it suggests his account needs to be modified. The first key characteristic of fictions that Vaihinger identifies, namely, their contradiction with reality, is maintained. The Rydberg electron in an atom is simply not following one of these classical closed orbit trajectories. In this sense, the fiction of closed orbits does involve a contradiction with reality. Vaihinger's second criterion, that the fiction is introduced only as a scaffolding to be eliminated, seems to require some modification. There is a straightforward sense in which classical trajectories in atoms have *already* been eliminated and replaced by the correct probabilistic description in terms of modern quantum mechanics. Indeed, we saw in the brief history at the beginning of section 3 that Bohr had initially introduced classical trajectories in atoms as an hypothesis; by the early 1920s it had been transformed from an hypothesis to a useful fiction, and by the end of that decade the fiction, which had indeed been a fruitful scaffolding, had been eliminated in favor of what we would call the true description of the behavior of electrons in atoms. Yet even some 30 years after these electron trajectories were eliminated and replaced, the fiction was reintroduced. The justification for the reintroduction of fictional electron trajectories was precisely their great fertility and explanatory power. Vaihinger's account does not seem to recognize this possibility, that some fictions may remain a part of the toolbox of science even after the true description has been found. Hence not all fictions are introduced to be eliminated.

Vaihinger's third key characteristic of fictions, namely, that the scientists expressly recognize that the fiction is just a fiction, is certainly maintained. At no point do semiclassical theorists, such as Delos and Kleppner, really think that the electron in the atom is following such a trajectory. As I mentioned earlier, I think this third condition is an important factor in the legitimate use of fictions in science. Semiclassical theorists deploy closed orbit theory always with the express recognition that the trajectories being invoked are fictional. At no point are they rejecting modern quantum mechanics, which denies the existence of such trajectories. Nonetheless, simply recognizing the fictional status of a posit does not mean that all such fictional posits are on par with each other, as Vaihinger seems to suggest. Even if we restrict ourselves, as Vaihinger does in his fourth key characteristic, to those scientific fictions that are expedient, there are still some important distinctions to be made—it is not the case that even all *expedient* fictions are on par with each other. In particular, I have argued that we need to distinguish between those fictions that are mere calculational tools, and those fictions that carry some explanatory force.

Vaihinger denied that there could be such explanatory fictions because he believed that fictions could not generate real knowledge. Using the example of classical trajectories in quantum spectra, I have tried to show that this assumption is mistaken. Some fictions can capture in their fictional representation real patterns of structural dependencies in the world, and hence generate real knowledge and be genuinely explanatory. It is noteworthy that semiclassical approaches to quantum phenomena, such as closed orbit theory, are not primarily valued as calculational tools. Indeed, in some cases the semiclassical calculations are just as complicated as—if not more complicated than—their full quantum counterparts. Instead, semiclassical approaches are valued because they provide an unparalleled level of physical insight into the structure of the quantum phenomena—a level of understanding that is difficult to extract from the full quantum solutions, even in those rare cases where the quantum solutions are available.

I suggested that one the chief obstacles to admitting the explanatory power of fictions is the lack of a philosophical framework for understanding how it is that fictions could explain. As I showed earlier, both of the current orthodox accounts of scientific explanation rule out the possibility of explanatory fictions. In response I outlined a new account of scientific explanation, called structural model explanations, and argued that this account provides us with a way of making sense of how it is that some fictions—despite their ontological status—nonetheless deserve the epithet *explanatory*.

NOTES

1. The revival of interest in Vaihinger's account of fictions is largely owed to Arthur Fine (1993), reprinted in this volume.
2. More precisely, these four characteristics together define what Vaihinger calls "scientific semi-fictions"—the type of fiction most relevant to the discussion here.
3. Although Vaihinger sees a clear logical distinction between fiction and hypotheses, as well as noting that they lay out very different methodologies, he does recognize that in the actual history of science this distinction may be difficult to draw. There may be cases, for example, where the scientist making the assertion is unsure of whether it is to be properly thought of as a hypothesis or fiction when it is first introduced. In such a case Vaihinger notes that methodologically it is best to assume that it is a hypothesis, and in that way not block the road to verification by prematurely declaring it a fiction.
4. For an explication of Bohr's much misunderstood correspondence principle, see Bokulich (2008a, chap. 4, sec. 2).
5. The spectroscopic data and Zeeman effects that most of us are familiar with (including the anomalous and Paschen–Bach effects) take place in the regime in which the external magnetic field is relatively weak compared with the electrostatic Coulomb field of the atom. If, however, the magnetic field strength is increased so that it is comparable to the Coulomb field, then diverse new phenomena occur, collectively known as the quadratic Zeeman effect. The quadratic Zeeman effect is so named because the Hamiltonian of an atom such as hydrogen in a magnetic field has two terms involving the magnetic field, B: one that is linear in B and one quadratic in B (that is, it has B^2). For a sufficiently weak magnetic field, one can ignore the quadratic term in the Hamiltonian and only the linear term is important; if, however, the magnetic field is very strong, then the quadratic term cannot be neglected. Atoms in strong magnetic fields are often referred to as diamagnetic atoms.
6. Even these higher resolution experiments were still of finite resolution, not resolving individual energy levels. What Figure 6.2 shows is the average absorption curve as a function of energy.

7. 'v-type' just refers to the clearly observable strong peaks in the Fourier-transformed spectra, with each peak being labeled with an integer, v.

8. For a more technical discussion of the closed orbit theory explanation of the spectra see Bokulich (in 2008a, Chapter 5, section 4).

9. See Granger (2001, Chapter 1) for a review.

10. See Du and Delos (1988) for a more detailed explanation, as well as a picture of a typical chaotic trajectory of Rydberg electron in a Coulomb and diamagnetic field.

11. There are, of course, consistent interpretations of quantum mechanics, such as Bohm's hidden variable theory, in which electrons do follow definite trajectories. Typically, however, these Bohmian trajectories are not the trajectories of classical mechanics. As an empirically equivalent formulation of quantum mechanics, Bohm's theory would nonetheless be able to account for any experimental results just as the standard interpretation. For those who are interested in Bohmian mechanics, a Bohmian approach to these diamagnetic Rydberg spectra has been carried out by Alexandre Matzkin, who concludes, "*Individual* BB [deBroglie–Bohm] trajectories do not possess these periodicities and cannot account for the quantum recurrences. These recurrences can however be explained by BB theory by considering the *ensemble* of trajectories . . . although none of the trajectories of the ensemble are periodic, rendering unclear the dynamical origin of the classical periodicities" (Matzkin, 2006, p. 1; emphasis original).

12. This definition of structural explanation is my own, and is not exactly identical to the definitions given by other defenders of structural explanations, such as Railton, Hughes, and Rob Clifton. Nonetheless, I think this definition better describes the concrete examples of structural explanations that these philosophers give, and is preferable given the notion of 'model' being used here. For further discussion of structural explanations and some examples see Bokulich (2008a, Chapter 6, section 5).

13. The technical details of this justificatory step are reviewed in Bokulich (2008a, equations 5.9–5.14).

7 Fictions, Representations, and Reality

Margaret Morrison

1. INTRODUCTION

There are many ways in which we use unrealistic representations for modeling the physical world. In some cases we construct models that we know to be false—but not false in the sense that they involve idealization or abstraction from real properties or situations; most models do this. Instead, they are considered false because they describe a situation that cannot, no matter how many corrections are added, be physically true of the phenomenon in question. Maxwell's ether models are a case in point. No one understood or believed that the structure of the ether consisted of idle wheels and rotating vortices, yet those types of models were the foundation of Maxwell's first derivation of the electromagnetic field equations.

Other instances of model building involve mathematical abstractions that are also not accurate representations of physical phenomena. For example, in his work on population genetics, R. A. Fisher (1918, 1922) assumed an analogy between populations of genes and the way that statistical mechanics models populations of molecules in a gas. These populations contained an infinite number of genes that act independently of each other. His inspiration was the velocity distribution law, which gave results about pressure (among other things) from highly idealized assumptions about the molecular structure of a gas. This kind of abstraction (assuming infinite populations of genes) was crucial in enabling Fisher to show that selection indeed operated in Mendelian populations.[1] In situations like this where we have mathematical abstractions that are *necessary* for arriving at a certain result there is no question of relaxing or correcting the assumptions in the way we de-idealize cases like frictionless planes and so on; the abstractions are what make the model work.

Another type of model or modeling assumption(s) that also resists corrections is the type used to treat a specific kind of problem. Deviations from those particular situations typically do not involve corrections to the model's assumptions but the introduction of a new model that describes the situation somewhat differently. For example, the typical Fisher–Wright model in modern population genetics assumes that generations are discrete—they do not overlap. However, if we want to examine changes in allele frequencies when we have overlapping generations, we don't simply add corrections or parameters to the discrete generation models; instead a different model is used (Moran model) for which parallels to the Fisher–Wright models can be drawn. A similar situation occurs in the case of nuclear models. The kinds of assumptions about nuclear structure required by the liquid drop model for explaining and predicting fission are very different from those contained in the shell model for explaining magic numbers.[2] These kinds of models may be fictional in a way that is similar to Maxwell's ether models, but the important point is that they are not open to the addition of correction factors or de-idealization.

Finally, we have the relatively straightforward cases of idealization where a law or a model idealizes or leaves out a particular property but allows for the addition of correction factors that bring the model system closer (in representational terms) to the physical system being modelled or described. The Hardy–Weinberg law is a good example of a case where violations of certain conditions, like random mating, may mean that the law fails to apply, but in other situations, depending on the type and degree of deviation from idealized conditions (e.g., no mutation), the law may continue to furnish reasonably accurate predictions. In other words, the population will continue to have Hardy-Weinberg proportions in each generation but the allele frequencies will change with that condition. The simple pendulum is a familiar example where we know how to add correction factors that bring the model closer to concrete phenomena. The key in each of these cases is that we know how to manipulate the idealizations to get the outcomes we want.

Because each of the cases I have just mentioned has a different structure, the important question is whether they exemplify anything *different* about the way unrealistic representations yield reliable knowledge.[3] I see this not as a logical problem of deriving true conclusions from false premises but rather an epistemic one that deals with the way false representations transmit information about concrete cases.[4] The latter is a problem that in some sense pervades many cases of knowledge acquisition to a greater or lesser extent—think of the use of metaphors in transmitting information. It is tempting to classify all of the examples just given as instances of fictional representation and then ask how fictions give us knowledge of real-world situations. I want to argue that this would be a mistake. The language of fictions is at once too broad and too narrow. Although it encompasses the fact that none of these representations is realistic, it fails to capture specifics of the relation that certain kinds of model-representations have to real systems. I claim that this is because the processes involved in both abstraction and idealization are not typically the same as those involved in constructing fictional models/representations. Introducing a mathematical abstraction that is necessary for obtaining certain results involves a different type of activity than constructing a model you know to be false in order to see whether certain analogies or similarities can be established. To simply classify all forms of nonaccurate description as fictions is to ignore the different ways scientific representation is linked with explanation and understanding.

I want to address the problem of unrealistic representation by introducing a finer grained distinction that uses the notion of fictional representation to refer only to the kind of models that fall into the category occupied by Maxwell's ether models. My reasons for doing so reflect the practice behind the construction of these types of models. Fictional models are deliberately intended as imaginary accounts whose physical similarity to the target system is not immediately obvious. Instead, one needs to examine the specific details of the model in order to establish the appropriate kinds of relations. Contrast this with the use of idealization where we have conditions that have been deliberately omitted or idealized (frictionless planes) in order to facilitate calculation or to illustrate a general principle for a simple case. Here we usually know immediately what the purported relation to the target system actually is.

In keeping with my rather narrow account of fictional representation, I want to also suggest a way of thinking about abstraction that differentiates it from idealization in the following way: Where idealization distorts or omits properties that are often not necessary for the problem at hand, abstraction (typically mathematical in nature) *introduces* a specific type of representation that is not amenable to correction and is necessary for explanation/prediction of the target system. What is crucial about abstraction, characterized in this way, is that it highlights the fact that the process is not simply one of adding back and taking away as characterized in the literature; instead it shows how certain kinds of mathematical representations are essential for explaining/predicting concrete phenomena.[5] Once again, my aim in distinguishing these different kinds of representation is to call attention to the fact that the notion of a 'fiction' is not sufficiently rich to capture the various ways that mathematical abstraction and idealization function in explanation and prediction.

So, what is it about the *structure* of each of these types of representation that makes them successful in a particular context? As I mentioned earlier, this isn't a logical problem in that we aren't so much concerned with the informational content or argument structure (that we can get true information from false premises) but rather with the features of the *representation* that produces knowledge. For example, what aspects of populations characterized by the Hardy–Weinberg law account for the latter's success in predicting genotype frequencies?[6] Similarly, what was the essential feature in Maxwell's ether model that led to the derivation of the field equations? So, there are really two interrelated issues here: (1) How do fictional representations provide reliable information, and (2) what is essentially different about the way that fictional models as opposed to abstraction and idealization accomplish this? My intuition is that although I can draw some general conclusions about the differences between fictions, abstractions, and idealizations, the answer to (1) will be a highly context-specific affair. That is, the way a fictional model produces information will depend largely on the nature of the model itself and what we want it to do. That said, it is important to recognize that there are stories to tell about how knowledge is produced in these cases that go beyond the simple appeal to heuristic power.

Let me begin by discussing some general issues related to fictions and falsity and then go on to examine a case of a fictional model, an idealized law, and a mathematical abstraction in order to illustrate some of the differences among them and attempt to answer some of the questions just raised.

2. FABLES, FICTIONS, AND FACTS

One of the things that is especially puzzling about fictional scientific representation is the relationship it bears to, say, literary fictions. Although the latter describe worlds that we know are not real, the intention, at least in more meritorious works of fiction, is often to shed light on various aspects of our life in the real world. To that extent some kind of parallel relationship exists between the two worlds that makes the fictional one capable of "touching", in some sense, the real one. But what does this "touching" consist of? There are aspects of the fictional world that we take to be representative of the real world but only because we can draw certain parallels or assume certain similarity relations hold. For example, many of the relationships described in the novels of Simone de Beauvoir can be easily assimilated to her own experiences and life with Sartre. In other words, even though the characters are not real, the dynamic that exists between them may be an accurate depiction of the dynamic between real individuals.

But what about scientific fictions? There too we have a relationship between the real world and the world described by our models. We also want to understand certain features of those models as making some type of *realistic* claim about the world. So, although we sometimes trade in analogies or metaphors, the goal is to represent the world in as realistic a way as we are able or in as realistic a way that will facilitate calculation or understanding of some aspect of the target system. Sometimes the relation between the fictional and the real is understood in terms of the abstract and the concrete, where abstract entities or concepts are understood as fictional versions of concrete realistic entities. And, unlike the literary case, we don't always understand the model, as a whole, to be a fictional entity; sometimes there are aspects of the model that are intended as realistic representations of the system we are interested in. The question, of course, is how exactly these models transmit reliable information about physical systems.

Nancy Cartwright (1999a), in a paper called "Fables and Models," draws on the ideas of Lessing about the relationship between fables and morals as a way of shedding light on the relation between the abstract and the concrete. Lessing sees a fable as a way of providing a graspable, intuitive content for abstract symbolic judgements (the moral). Fables are like fictions—they are stories that allegedly tell us something about the world we inhabit. The interesting thing about Lessing's fables is that they aren't allegories. Why is that? Because allegories function in terms of similarity relations; they don't have literal meaning and say something similar to what they seem to say. But, as Cartwright notes, for Lessing similarity is not the right relation to focus on. The relation between the moral and the fable is that of the general to the more specific; indeed, it is a misusage to say that the special bears a similarity to the general or that the individual has a similarity with its type. In more concrete terms, Lessing's account of the relation between the fable and the moral is between the abstract symbolic claim and its more concrete manifestation.

Cartwright claims that this mirrors what is going on in physics. We use abstract concepts that need "fitting out" in a particular way using more concrete models; the laws of physics are like the morals and the models like the fables. It is because force is an abstract concept that it can only "exist in particular mechanical models" (1999a, p. 46). What she concludes is that the laws are true only of objects in the model in the way that the morals are true of their fables; so, continuing on with Lessing's analysis, we would say that the model is an instance of the law. In that sense the model is less abstract than the law; but what about the relation of the model to the world? Cartwright says she is inclined to think that even when models fit "they do not fit very exactly" (1999a, p. 48). This provides a context in which to understand Cartwright's claim about how laws can be both false and have broad applicability. They are literally false about the world, yet the concrete models that instantiate them are what constrain their application, and it is because the models also don't have an exact fit that they have such broad applicability.

But what exactly does this view entail when we move from the model to the world? If, as it seems to suggest, we can only talk about laws in the context of the fictional world described by the model, then how do we connect the fictional model with the real world that we are interested in explaining/predicting/describing? Should we understand reality as an instance of the model in the way that the model is an instance of the law, or do we need to invoke the similarity relation as a way of understanding why the model works in the way that it does? It isn't clear to me how talking in terms of instances or appealing to the general/specific relation tells us much here. When we need to know the features of the model that are instantiated in the world, we are essentially asking how the model is similar to the world. In other words, when we want to know some *facts* about the world, we need to move beyond the relation of law to model to one of model and world. Understanding the model as a concrete instantiation of a law doesn't guarantee that the model bears any similarity to the physical world. Maxwell's initial ether model was an instantiation of laws of hydrodynamics, yet it was a highly fictional representation of the ether/electromagnetic field. So, if our laws only say something about the world in virtue of the relation that the model bears to the world, the question becomes one of determining what, exactly, fictional models say about the world and the way in which they do it. The problem, however, is that if all models are fictions then we seem forced to conclude that science provides information about the world in the same way that novels do.

This characterization seems unhelpful, primarily because it fails to do justice to the way models are used in unearthing aspects of the world we want to understand. Put differently, we need to know the variety of ways models can represent the world if we are to have faith in those representations as sources of knowledge. To say force is an abstract concept that exists only in models leaves us with no insight about how to deal with physical forces that we encounter in the world. And, as I mentioned earlier, to characterize models generally as fictions doesn't tell us much either. We need a finer grained distinction that will capture the various types of unrealistic representations that are used in model construction and how those representations function in an explanatory or predictive way. Fictional representation is just one type. In order to see how we might make sense of the idea that fictional models can provide us with information about the world, and how we can retrieve information directly from idealized laws/models and mathematical abstractions, let us look at some examples of how this happens.

3. FICTIONAL MECHANISMS YIELD ACCURATE PREDICTIONS: HOW THE MODEL PROVIDES INFORMATION

In the various stages of development of the electromagnetic theory, Maxwell used a variety of tools that included different forms of a fictional ether model as well as physical analogies. Each of these played an important role in developing both mathematical and physical ideas that were crucial to the formulation and conceptual understanding of field theory. In order to appreciate exactly how a field theory emerged from these fictional representations, we need to start with Maxwell's 1856 representation of Faraday's electromagnetic theory in what he called a "mathematically precise yet visualizable form" (Maxwell, 1856; hereafter *FL*).[7] The main idea in Faraday's account was that the seat of electromagnetic phenomena was in the spaces surrounding wires and magnets, not in the objects themselves. He used iron filings to visualize the patterns of these forces in space, referring to the spatial distribution as lines of force that constituted a kind of field. Electrical charges were conceived as epiphenomena that were manifestations of the termination points of the lines of force, and as such, they had no independent existence. On this picture the field was primary with charges and currents emerging from it.

The method Maxwell employed involved both mathematical and physical analogies between stationary fields and the motion of an incompressible fluid that flowed through tubes (where the lines of force are represented by the tubes). Using the formal equivalence between the equations of heat flow and action at a distance, Maxwell substituted the flow of the ideal fluid for the distant action. Although the pressure in the tubes varied inversely as the distance from the source, the crucial difference was that the energy of the system was in the tubes rather than being transmitted at a distance. The direction of the tubes indicated the direction of the fluid in the way that the lines of force indicated the direction and intensity of a current. Both the tubes and the lines of force satisfied the same partial differential equations. The purpose of the analogy was to illustrate the mathematical similarity of the laws, and although the fluid was a purely fictional entity it provided a visual representation of this new field theoretic approach to electromagnetism.

What Maxwell's analogy did was furnish what he termed a physical "conception" for Faraday's lines of force; a conception that involved a fictional representation, yet provided a mathematical account of electromagnetic phenomena as envisioned on this field theoretic picture. The method of physical analogy, as Maxwell referred to it, marked the beginning of what he saw as progressive stages of development in theory construction. Physical analogy was intended as a middle ground between a purely mathematical formula and a physical hypothesis. It was important as a visual representation because it enabled one to see electromagnetic phenomena in a new way. Although the analogy did provide a model (in some sense), it was merely a descriptive account of the distribution of the lines in space with no mechanism for understanding the forces of attraction and repulsion between magnetic poles.

A physical account of how the behavior of magnetic lines could give rise to magnetic forces was further developed in a paper entitled "On Physical Lines of Force" (Maxwell, 1861–1862; hereafter *PL*). The paper marked the beginnings of his famous ether model, which described the magnetic field in terms of the rotation of the ether around the lines of force. The idea of a rotating ether was first put forward by Kelvin, who explained the Faraday effect (the rotation of the plane of polarized light by magnets) as a result of the rotation of molecular vortices in a fluid ether. In order to allow for the rotation of the adjacent vortices, the forces that caused the motion of the medium and the occurrence of electric currents, the model consisted of layers of rolling particles between the vortices. The forces exerted by the vortices on these particles were the cause of electromotive force, with changes in their motion corresponding to electromagnetic induction. That Maxwell thought of this as a purely fictional representation is obvious from the following quotation:

> The conception of a particle having its motion connected with that of a vortex by perfect rolling contact may appear somewhat awkward. I do not bring it forward as a connexion existing in nature, or even as that which one would willingly assent to as an electrical hypothesis. It is,

however, a mode of connexion which is mechanically conceivable, and easily investigated. (Maxwell, 1965, vol. 1, p. 486)

One of the problems with the fluid vortex model was how to mechanically explain the transmission of rotation from the exterior to the interior parts of each cell. How could a fluid surface exert tangential forces on the particles? In order to remedy this and extend the model to electrostatics Maxwell developed an elastic solid model made up of spherical cells endowed with elasticity which made the medium capable of sustaining elastic waves. In addition, the wave theory of light was based on the notion of an elastic medium that could account for transverse vibrations; hence it was quite possible that the electromagnetic medium might possess the same property. The fluid vortices had uniform angular velocity and rotated as a rigid sphere, but elastic vortices would produce the deformations and displacements in the medium that needed to be incorporated. With this elasticized medium Maxwell now needed to explain the condition of a body with respect to the surrounding medium when it is said to be charged with electricity and to account for the forces acting between electrical bodies.

According to the Faraday picture, electric lines of force were primary and electric charge was simply a manifestation of the terminating points on the lines of force. If charge was to be identified with the accumulation of particles in some portion of the medium, then it was necessary to have some way of representing that. In other words, how is it possible to represent charge as existing in a field? In the case of a charged capacitor with dielectric material between the plates, the dielectric material itself was

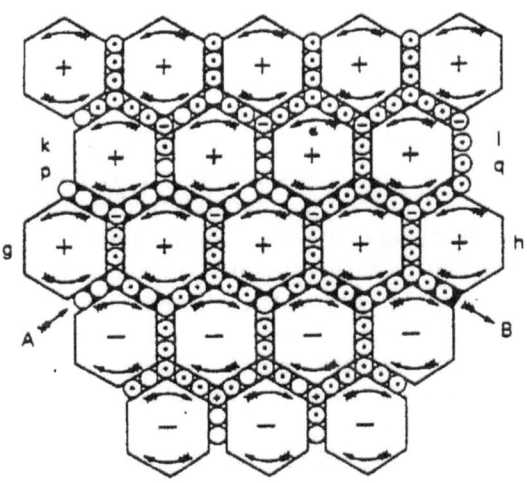

Figure 7.1 Maxwell's vortex ether model. AB is a current of electricity, with the large spaces representing the vortices and the smaller circles the idle wheels.

the primary seat of the "inductive" state and the plates served merely as bounding surfaces where the chain of polarized particles was terminated. So, what is taken to be the charge on a conductor is nothing but the apparent surface charge of the adjacent dielectric medium.

The elastic vortices that constituted the ether or medium were separated by electric particles whose action on the vortex cells resulted in a type of distortion. In other words, the effect of an electromotive force (the tangential force with which the particles are pressed by the matter of the cells) is represented as a distortion of the cells caused by a change in position of the electric particles. That, in turn, gave rise to an elastic force that set off a chain reaction. Maxwell saw the cell distortion as a displacement of electricity within each molecule, with the total effect over the entire medium producing a "general displacement of electricity in a given direction" (1965, vol. 1, p. 491). Understood literally, the notion of displacement meant that the elements of the dielectric had changed positions. And, because changes in displacement involved a motion of electricity Maxwell argued that they should be "treated as" currents in the positive or negative direction according to whether displacement was increasing or diminishing. Displacement also served as a model for dielectric polarization—electromotive force was responsible for distorting the cells and its action on the dielectric produced a state of polarization. From this we can get some sense of just how complex the model really was and how, despite its fictional status, its various intricacies provided a representation of important features of the electromagnetic field. But most important was how this account of the displacement current(s) furnished the appropriate mathematical representation that would give rise to the field equations.

In the original version of the ether model displacement was calculated only in terms of the rotation of the vortices without any distortion, but in the elastic model displacement made an additional contribution to the electric current. In order to show how the transmission of electricity was possible in a medium, a modification of Ampere's law was required in order to generalize it to the case of open circuits. What is important here, however, is not the modification of Ampere's law per se but rather the way in which the model informed that modification. If we think for a moment about what Maxwell was trying to achieve, namely, a field theoretic representation of electromagnetism, then it becomes obvious that some way of treating open circuits and representing charges and currents as emerging from the field rather than material sources is crucial.

To achieve that end two important elements were required. The first concerns the motion of idle wheels that represented electricity as governed by Ampere's law relating electric flux and magnetic intensity (curl $\mathbf{H} = 4\varpi\mathbf{J}$, where $\mathbf{H}$ is the magnetic field and $\mathbf{J}$ is the electric-current density). A consequence of that law was that it failed to provide a mechanism for the accumulation of charge because it applied only in the case of closed currents. Consequently a term $\partial\mathbf{D}/\partial t$ had to be added to the current so that it

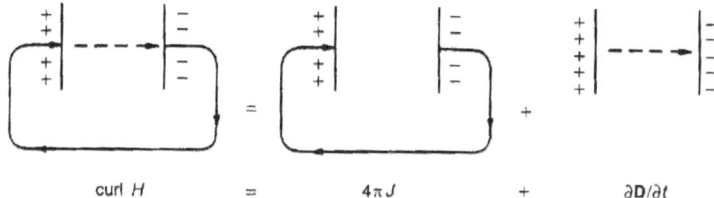

$$\text{curl } H \quad = \quad 4\pi J \quad + \quad \partial D/\partial t$$

Figure 7.2 The displacement term $\partial D/\partial t$ modified the original Ampere law where $D = (1/c^2)E + 4\varpi P$ (the polarization vector).

was no longer circuital. As we saw earlier, and from the diagram in Figure 7.2, the dielectric between the coatings of a condenser fulfilled that need and was seen as the origin of the displacement current. The force of that current was seen as proportional to the rate of increase of the electric force in the dielectric and hence produced the same magnetic effects as a true current. Hence the charged current could be "regarded as" flowing in a closed circuit.[8] The modified term in Ampere's law took the value zero for the case of steady currents flowing in closed circuits and nonzero values for open circuits, thereby giving definite predictions for their magnetic effects.

But, as I noted earlier, there was a second feature concerning the account of displacement, namely, how the representation of electric current qua elastic *restoring* force (a crucial feature of the model) was used to represent open circuits. According to the mechanics of the model, the rotations push the small particles along giving rise to a current of magnitude $1/4\varpi$ **curl H** while the elastic distortions move the particles in the direction of the distortion, adding another contribution $\partial D/\partial t$ to the current.[9] Maxwell had linked the equation describing displacement ($R = -4\varpi E^2 h$) with the ether's elasticity (modeled on Hooke's law) and also with an electrical equation representing the flow of charge produced by electromotive force. Hence R was interpreted as both an electromotive force in the direction of displacement and an elastic restoring force in the opposite direction. Similarly, the dielectric constant E is both an elastic coefficient and an electric constant, and h represents both charge per unit area and linear displacement. As a result, the equation served as a kind of bridge between the mechanical and electrical parts of the model. The electrical part included the introduction of the displacement current, the calculation of a numerical value for E (a coefficient that depended on the nature of the dielectric), and the derivation of the equations describing the quantity of displacement of an electric current per unit area. The mechanical part required that E represent an elastic force capable of altering the structure of the ether. The point that concerns us here, though, is exactly how the mechanical features of the model gave rise to electrical effects: that is, the relation between the mechanical distortion of the ether and the displacement current.

The answer lies in seeing how Maxwell's model represented the primacy of the field. As Seigel (1991, p. 99) points out, in Maxwell's equation of electric currents the **curl H** term (the magnetic field) appears on the right side with the electric current **J** on the left as the quantity to be calculated.

$$\mathbf{J} = 1/4\varpi \ (\textbf{curl H}—1/c^2 \ \partial E/\partial t)$$

To illustrate how this works mechanically, consider the following example: If we take a charging capacitor there is a growing electric field pointing from positive to negative in the space between the plates. This is the current owing to the solenoidal, closed loop **curl H** term. However, associated with this field there is a reverse polarization (Figure 7.3) due to the elastic deformation of the vortices acting on the particles. This gives rise to a reverse current between the plates that cancels the **curl H** term. This is because it is negative and points toward the positive plate as the capacitor is charging. The solenoidal term is incapable of producing accumulations of charge, so it is the reverse polarization that gives rise to charge on the capacitor plates and not vice versa. It is the constraint on the motion of the particles that reacts back as a constraint on the motion of the vortices that drives the elastic distortion of the vortices in the opposite direction. As a result, charge builds up through the progressive distortion of the medium. This elastic distortion is accompanied by a pattern of elastic restoring forces that correspond to the electric field *E*. Without this there would be no charge because it is responsible for relaxing the solenoidal property of the electric current.

We can now state the relation between charge and the field: Charge is the center of elastic deformation that gives rise to a pattern of electromotive forces, which constitutes the field—the energy of deformation is the electric field energy. Put simply, it is the fields that give rise to charges and currents. The magnetic field gives rise to the solenoidal in both the wire and space between the plates and the changing electric field gives rise to

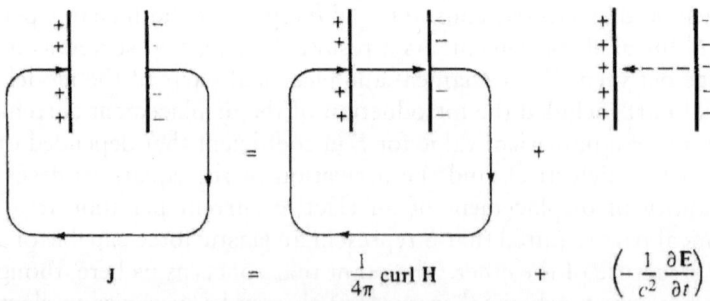

Figure 7.3 The reverse (polarization) current term has a negative sign, which cancels the **curl H** term between the plates.

a reverse current in the space between the plates. The sum of these two currents yields the conduction current **J**—an open circuit for the true current.

What this shows is how a completely fictional model provided an account of how electromagnetic forces could be produced in a mechanical ether. But, more significantly, it furnished the all important field theoretic picture that would become the basis for the modern account we have today. Once the displacement current secured the basis for a field theory, it was in some sense a rather small step to construct the *electromechanical* theory of light. I say electromechanical because this particular model didn't identify light with electromagnetic waves; rather, it was hypothesized that both originated from the same medium or field. Instead of approaching the problem as a mathematical one and obtaining solutions for the equations for the **E** or **H** fields in the form of transverse electromagnetic waves, the model occupied center stage with the mathematical account emerging from the model.

Very briefly, the rest of the 1861–1862 story goes something like this.[10] The velocity of propagation (V) of transverse torsion waves in the medium was given by the torsion modulus m, an elastic constant that controls the strength of the electric forces, divided by the mass density of the medium ρ_M which controls the strength of magnetic forces. Maxwell set the values of these parameters through a chain of linkages between the mechanical, electrical, and magnetic aspects of the model, which resulted in $V = c$ where c is the ratio of units—a measure of the relative strengths of electrostatic forces and electromagnetic forces. This ratio of electric to magnetic units depended on a quantity that had the dimensions of a velocity. There were five different methods for determining that velocity and using these experimental results Maxwell obtained a value for c that was very close to the velocity of light. Consequently the velocity of waves in the magnetoelectric medium was also roughly equivalent to the velocity of light.[11] But, and this is the important point, the rationale for setting the parameters to the values Maxwell chose, together with the equivalence between V and the ratio of units, followed directly from the mechanical and electromagnetic connections that emerged from the model. In some sense the model, like an experimental novel, had played out its own story; the succession of facts depended on the constraints imposed by the phenomena themselves via their place in the model.[12]

The problem of course was that no one, especially Maxwell, thought this model was anything but an elaborate fiction. But because the numerical relations between the optical and electromagnetic phenomena were too spectacular to ignore, he needed some way of showing that the existence of electromagnetic waves propagating with velocity c followed from electrical equations themselves, divorced from the mechanical underpinnings of the model. The answer came in 1865 via a purely dynamical theory and the introduction of electromagnetic variables into the equations of dynamics. However, with the abandonment of the mechanical model he also, and perhaps most importantly, needed a different justification for introducing displacement into his equations, for without it there was literally no field theory.

A number of assumptions made this possible. Based on facts about light he assumed that space or the ether must be capable of storing up energy, and from the forms of the equations of electricity and magnetism he showed the forms that energy must have if it is disseminated through space. If the medium is subject to dynamical principles then it will be governed by Lagrange's equations, and if we apply dynamics to electromagnetism we can substitute magnetic and electric energy for kinetic and potential energy. According to the Faraday–Mossotti theory of dielectrics, when a potential difference is applied across a dielectric it becomes polarized—molecules are positive at one end and negative at the other. This entails not only that electric energy is stored in the dielectric but that a transient electric current flows in it. If one assumed that space can store up energy it could also become polarized by a potential difference, and a changing potential difference would thereby produce a changing current (displacement) associated with a changing magnetic field. Once the displacement current was introduced Maxwell was able to deduce the properties of electromagnetic waves that accounted for light.

All of this was done in a paper entitled "A Dynamical Theory of the Electromagnetic Field." Of course, this is not to say that the new "dynamical theory" was without problems. Because there was no mechanical model it became difficult to conceive how exactly displacement operated. If electricity was being displaced, how did this occur? The problem was magnified because charge was interpreted as a discontinuity in displacement. And although displacement appears in the fundamental equations, divorced from its mechanical foundation, electromagnetism takes on the appearance of a phenomenological theory.[13] The continuation of this story and the efforts to unite mechanics and electromagnetism is both long and rather complicated.[14] Although it has interesting implications for debates about the relationship between fictional representations, mathematical representations, and concrete knowledge, those are not issues I can adequately deal with here. Instead I want to conclude this section by focusing on the problem I raised at the beginning, namely, what specific features can we isolate as playing a role in the transmission of information from fictional models?

What is the sense of representation that is important here and how does it emerge from the fictional model? What is especially significant is that the development of the field equations did not proceed through the introduction of a new term called the displacement current as a mathematical modification to Ampere's law. Instead, what I tried to show is how the foundation for electromagnetism emerged from the molecular vortex model and was in fact determined by it. But the important issue here is not that Maxwell was capable of deriving a set of field equations from a false model, but rather what it was about the model that underscored the applicability of the equations. Put differently, what was the conceptual or theoretical content represented in the model that formed the basis for his electromagnetic worldview?

The answer centers, of course, on the displacement current. Maxwell knew that the ether or field did not consist of rotating vortices and idle wheels, but he also knew that in order to represent the Faraday picture of electromagnetism there had to be some account of how electricity could travel in free space and charge could be build up without material bodies. Consequently, he needed a way of representing this mechanically if he was to derive a set of equations applicable to this new field theoretic picture. The mechanical model became the focal point for an understanding of how charges and currents could be understood as field theoretic phenomena and the formulation of a mathematical account of those processes. Part of that account involved a modification of the relation between between Ampere's law, Coulomb's law, and the equation of continuity for open circuits, something that was indicated by the structure of the model. Once the basic mechanical features of the model were in place they constrained both the physical and mathematical descriptions of electromagnetic forces. In the same way that character development in a novel determines, to some extent, how the story will play out, the features of the model restrict the way that certain physical relations can be represented. By the time the *Treatise on Electricity and Magnetism* was completed in 1873, displacement had taken on the role of a primary current responsible for the transmission of electricity through space. Although the mature theory was an extension of the dynamical approach, neither would have been possible without the fictional model that provided a physical conception of how an electromagnetic field theory might be possible.

What this extended discussion hopefully shows is not only how certain information emerged from the fictional model but also the need for examining the model in detail in order to show exactly how this happens. Although I have only addressed one example here, the point I want to make is a general one concerning fictional models. To say that fictional models are important sources of knowledge in virtue of a particular kind of similarity that they bear to concrete cases or systems is to say virtually nothing about how they do that. Instead, what is required is a careful analysis of the model itself to uncover the kind of information it yields and the ways in which that information can be used to develop physical hypotheses. There are various ways that fictional models may be able to accomplish this, but each one will do so in a way that is specific to that particular model. The situation is radically different from the case of idealization where an analysis of the methods employed for both the idealizing process and its corrections will typically cover a variety of different cases. Consequently, idealization becomes a relatively easy category to define, but it still presents some challenges for uncovering how idealized models relate to real systems.

4. FROM THE IDEAL TO THE CONCRETE

So, how exactly does the fictional case differ from an idealized law/model where there are specific unrealistic conditions required for the law to hold, conditions that are false but nevertheless capable of approximating real physical systems? The Hardy–Weinberg law is a simple example of a fundamental law that makes assumptions not realized in any natural population. In what sense do we want to say that this law provides us with information about these populations? One of the things that renders laws explanatory, as highlighted by the D-N model, is the fact that they are general enough to apply to a diverse number of phenomena. In other words, they enable us to understand specific features of phenomena as similar in certain respects; for example, universal gravitation shows that both terrestrial and celestial bodies obey an inverse square force law. Nancy Cartwright (1983) claims that this generality is a reason for thinking fundamental laws like these are false; their generality results in their being unable to fully describe the situations they reportedly cover, or they deliberately omit aspects of the situation that are not relevant for the calculation at hand. In that sense they don't accurately describe concrete situations and are true only of objects in our models. The problem then is how could they possibly provide knowledge of concrete physical systems, or in this case, populations?

Part of Cartwright's reason for claiming that covering laws are false is to contrast them with phenomenological laws (or models) that supposedly do give us more accurate descriptions of the physical world. However, what the Hardy–Weinberg law shows is that embedded in what Cartwright would call a 'false' law is a great deal of accurate information about biological populations, information that was crucial in the synthesis of Mendelian heredity and Darwinian natural selection. To that extent it serves as an example of how mathematical idealization (and abstraction) can enhance our understanding far beyond simple predictive capabilities.[15] As we shall see later in this chapter, the Hardy–Weinberg law enables us to understand fundamental features of heredity and variation by establishing a mathematical relation between allele and genotype frequencies that embodies the very gene conserving structure that is the essential feature of Mendelism. What is important for my purposes here is to show why the unrealistic nature of its assumptions does not affect the significance of either the conclusions it provides or the information implicit in its formulation. Moreover, it is a nice example of the differences I want to highlight between idealization and the account of abstraction I described in the introduction.

The Hardy–Weinberg law is often described as a consequence of Mendel's law of segregation, or a generalization of Mendel's laws as applied to populations. It relates allele or gene frequencies to genotype frequencies and states that in an infinite, random mating population in the absence of external factors such as mutation, selection, drift, and migration, one generation of random mating will produce a distribution of genotypes that is a function solely of allele frequencies. Moreover, this distribution does not change over subsequent generations, provided all conditions are held constant. In other words, if we have a pair of alleles Aa at a particular gene locus and the initial ratio of A to a is p to q, then for every succeeding generation the ratio will be p to q. Regardless of the distribution of genotypes in the initial generation the distribution for all succeeding generations will be

$$p^2A_1A_1 + 2pqA_1A_2 + q^2A_2A_2$$

where p^2 is just the probability of getting an A_1A_1 homozygote, which is the probability that the egg is A_1 times the probability that the sperm is A_1 (by the product rule for independent events). Both of these probabilities are p because in its simplest form the law assumes that the species is hermaphroditic. Because the heterozygote can be formed in two different ways the probability is $2pq$ (by the addition rule for mutually exclusive events). So, if you know the value for p then you know the frequencies of all three genotypes.

Because random mating does not change allele frequencies, all one needs to calculate the genotype frequencies after a round of random mating is the allele frequencies before random mating. In populations where each individual is either male or female with different allele frequencies it will take two generations to reach Hardy–Weinberg equilibrium. One can see then the relation between the stability of the frequencies and Mendel's law of segregation. With random cross-fertilization there is no disappearance of any class whatever in the offspring of the hybrids, and each class continues to be produced in the same proportion.[16]

But, and here is the important point, what is significant about the Hardy–Weinberg law is not so much the binomial form of the genotype frequency and the prediction of genotypes based on the stability of the population, but rather what the stability actually shows or presupposes. Despite the idealizing assumptions, the stability allows us to understand something about Mendelian populations that is significant for understanding heredity and variation. In other words, certain conditions must be present for the stability to be possible. Thus, the predictive success of the law is intimately connected with certain basic claims about genetic structure that are presupposed in its formulation. What the Hardy–Weinberg law says is that if no external forces act, then there is no intrinsic tendency for the variation caused by the three different genotypes that exist in a population to disappear. It also shows that because the distribution of genotype frequencies is independent of dominance, dominance alone cannot change genotype frequencies. In other words, there is no evidence that a dominant character will show a tendency to spread or a recessive one to die out. Instead, the genotype frequencies are maintained in constant proportions. The probabilistic genetic structure is conserved indefinitely; but should it be influenced by an outside force, such as mutation, the effect would be preserved in a new stable distribution in the succeeding generation.

This was crucial for understanding the problems with blending inheritance as advocated by the Darwinians, and to that extent the claim that the law is false in some sense misses the point if our concern is conveying information. Under blending inheritance, variation was thought to decrease rapidly with each successive generation, but Hardy–Weinberg shows that under a Mendelian scheme it is maintained. This pointed to yet another fundamental aspect of Mendelism, namely, the discontinuous nature of the gene, and why it was crucial for the preservation of variation required for selection. How was it possible for the genetic structure to be maintained over successive generations? The reason for the stability could be traced directly to the absence of fusion, which was indicative of a type of genetic structure that could conserve modification. This condition was explicitly presupposed in the way the law was formulated and how it functioned.[17] In that sense one can see the Hardy–Weinberg (H-W) law as the beginning of a completely new explanation of the role of mutation and selection and how they affect our understanding of evolution.

What I have focused on thus far has been the information about the nature of heredity that is embedded or presupposed in the structure of this law. But what about the so-called "falsity" of the assumptions under which it is alleged to hold? Although the "model population" specified by the H-W law bears little, if any, relation to actual human populations, we saw that the law was an important source of theoretical information. What happens when assumptions like random mating and infinite populations are replaced with more realistic assumptions true of actual populations, assumptions like assortative mating and small populations? Although there are many forms of nonrandom mating, what is crucial from the point of view of "unrealistic assumptions" is that in many cases it is possible to show that given a parental and daughter generation allele frequencies remain the same in each generation; hence genetic variation is maintained—the fundamental conclusion of the H-W law. However, because heterozygote frequency is less than that applying in random mating populations, the variation is in some sense cryptic; that is, you get the same allele frequencies but different genotype frequencies. But after one generation of random mating, the H-W genotype frequencies would be immediately restored. Similarly, in the case of infinite populations, once we relax this assumption we find that mean heterozygosity decreases very slowly with time as a result of the sampling drift implicit in the process. This slow loss can be understood as the stochastic analogue of the "variation-preserving" property of infinite populations described by H-W. Although a violation of these conditions destroys H-W equilibrium, we nevertheless learn some useful information about the population.

Appealing to the abstraction/idealization distinction I introduced at the beginning can further clarify our understanding of how deviations from the conditions or assumptions specified by the law affect its applicability. Essentially we can divide the assumptions associated with the Hardy–Weinberg law into two groups. The first involves assumptions that don't allow for relaxation without violating H-W equilibrium, such as infinite population size and random mating. The second includes the absence of selection, migration, and mutation. These assumptions affect allele frequencies but not random mating. For example, selection may be taking place in a population that is nevertheless breeding randomly. Violations of these latter conditions will not rule out H-W proportions; instead, the allele frequencies will change in accordance with the changing conditions. In other words, these conditions function as idealizations that may or may not hold but whose effect on the system can be straightforwardly calculated. Put differently, we can think of them as external factors that isolate basic features of a Mendelian system that allow us to understand how variation could be conserved.

Contrast that situation with the requirements of infinite populations and random mating. Infinite populations are crucial in that one must be able to rule out genetic drift, which is a change in gene frequencies that results from chance deviation from expected genotypic frequencies. That is, we must be able to determine that detected changes are not due to sampling errors. Although random mating seems like the kind of restriction that is typically violated, we can see how its violations affect gene frequencies: In the case of assortative mating there will be an increase in homozygosity for those genes involved in the trait that is preferential such as height or eye color. Traits such as blood type are typically randomly mated. Similarly, in the case of inbreeding there will be an increase in homozygosity for all genes. Because both of these assumptions are necessary for H-W equilibrium, they cannot, in general, be corrected for and in that sense are necessary features for the applicability of the law. In other words, they ought to be considered abstractions rather than idealizations because they describe situations that cannot approximate real-world situations through the addition of correction factors.[18]

My account of abstraction is somewhat different from that typically described in the literature. Although abstraction and idealization are sometimes conflated, Cartwright (1989) has distinguished them in the following way: Idealization is a process where one starts with a concrete object and then mentally rearranges some of its (inconvenient) features or properties. This enables us to write down a law describing its behavior in certain circumstances. In some cases it is possible to just omit factors that are irrelevant to the problem, but for the factors that are relevant they are sometimes given values that are not, strictly speaking, accurate but allow for ease of calculation. The idealizations presupposed by the ideal gas law are an example (infinitesimal size of molecules and absence of intermolecular forces). In these cases we sometimes know the degree to which the idealization is a departure from the real situation and if necessary its effect can be estimated.

Abstraction presents a different scenario; here what Cartwright calls the "relevant features" have been genuinely subtracted. The example she uses to illustrate the point is the comparison between a law and a model. Many of the effects studied in modern physics make a very small contribution to the total behavior of a system and hence the laws governing these systems do not really approximate, in any real sense, what happens in concrete cases. These laws are instances of abstractions. The model, on the other hand, takes the relevant factors and assigns them convenient values in order to facilitate calculation. Although the latter may be unrealistic in the sense that it gives an idealized representation of particular properties, it still makes contact with the world insofar as it includes properties that are relevant to the system's behavior. Abstract laws do not literally describe the behavior of real systems because (1) they subtract features in order to focus on a single set of properties or laws as if they were considered in isolation and (2) no amount of theory will ever allow us to complete the process of concretization. Laws that govern the laser abstract from its material manifestations to provide a general description that is "common to all, though not literally true of any" (1989, p. 211). By contrast, the assumption of infinitesimal size for molecules can be corrected to make the ideal case more realistic.

Although I think Cartwright is essentially right in her claim that we can never "concretize" abstractions, the question that interests me is why that is the case. In her discussion of abstraction Cartwright mentions Duhem, who is also concerned with the notion of abstraction in physics. Because physics needs to be precise, it can never fully capture the complexity that is characteristic of nature. Hence, the abstract mathematical representations used by physics do not describe reality but are better thought of as imaginary constructions—neither true nor false. Laws relate these symbols to each other and consequently are themselves symbolic. Although these symbolic laws will never touch reality, so to speak, they do involve approximations that are constantly undergoing modification due to increasing experimental knowledge (Duhem, 1977, p. 174). In that sense, Duhem's account of abstraction seems also to incorporate elements of what Cartwright would call idealization. It also seems clear from his account that the gap between symbolic laws and reality is due, essentially, to an epistemic problem that besets us in the quest for scientific knowledge. Laws are constantly being revised and rejected; consequently, we can never claim that they are true or false. In addition, because of the precise nature of physics we must represent reality in a simple and incomplete way in order to facilitate calculation. This describes both idealization and abstraction, depending on how one chooses to simplify.

What Duhem's view captures is a philosophical problem that focuses on the gap between reality and our representation of it. As we saw earlier, the account of abstraction I am concerned with has its basis in this gap as well, but is motivated by particular kinds of abstract representations that are required for dealing with certain kinds of systems. Put differently, both Duhem's and Cartwright's accounts of abstraction describe physics more generally and the problems that beset a realist interpretation of theories, laws, and models. Although those issues have some bearing on my point, my overall concern stems from the way that we are *constrained* to represent certain kinds of systems in a mathematically abstract way if we are to understand how they behave. Because this goes beyond problems of calculation to issues about explanation, 'abstraction as subtraction' is not a useful category for my purposes. In my account one of the things that makes abstraction especially interesting is that it is the mathematical representation that provides the foundation for understanding causal features of the system. The physics and mathematics are inextricably intertwined, making our very characterization of these systems mathematical abstractions.

5. BEYOND FICTIONS: THE NECESSITY OF ABSTRACTIONS

In the preceding discussion we saw how infinite populations were an important constraint in the operation of the Hardy–Weinberg law. They are necessary for eliminating the chance or random influences on gene frequencies from one generation to the next, something that is common in small populations. Deviations from infinite population size can, of course, be handled because the kinds of populations to which one applies these models are never infinite. But in moving to smaller populations one must recognize that as population size decreases the effects of drift will become more predominant, thereby making it difficult to determine whether particular features of the population are the result of drift or selection. In other words, we cannot determine whether the population is undergoing evolutionary change. In that sense infinite population size (along with random mating) is a necessary condition for H-W equilibrium to be maintained and for determining how deviations from it (selection, mutation, etc.) are to be understood.

Other cases where abstract representations are crucial for understanding how the system in question behaves include phase transitions. There are a variety of physical systems that fall into this category: superconductivity, superfluidity, magnetism, crystallization, and several others. In each of these cases there is a spontaneous symmetry breaking associated with a phase transition that explains the occurrence of the superconducting or magnetic state of matter. Very briefly, the situation is as follows: In thermodynamics (TD), phase transitions are accounted for in terms of discontinuities in the thermodynamic potentials. However, once we move to statistical mechanics (SM) the equations of motion that govern these systems are analytic and hence do not exhibit singularities. As a result, there is no basis for explaining phase transitions in SM. In order to recover the TD explanation we need to introduced singularities into the equations and thus far the only way to do this is by assuming the number of particles in the system is infinite. That is, we need to invoke the "thermodynamic limit," $N \rightarrow \infty$, which assumes that the system is infinite in order to understand the behavior of a real, finite system. Note that the problem here isn't that the limit provides an easier route to the calculational features associated with understanding phase transitions; rather, the assumption that the system is infinite is *necessary* for the phase transitions to occur. In other words, we have a description of a physically unrealizable situation that is *required* to explain a physically realizable one (the occurrence of phase transitions).

Given this situation, how should we understand the relation between the abstract description and the concrete phenomena it supposedly explains? Although one might want to claim that within the *mathematical* framework of SM we can causally account for (explain) the occurrence of phase transitions by assuming the system is infinite, it is nevertheless tempting to conclude that this explanation does not help us to *physically* understand how the process takes place because the systems that SM deals with are all finite. Similar doubts have been expressed by Callender (2001) and Earman (2003), who argues against taking the thermodynamic limit as a legitimate form of idealization: "a sound principle of interpretation would seem to be that no effect can be counted as a genuine physical effect if it disappears when the idealizations are removed" (p. 21). Both claim that, we shouldn't assume phase transitions have been explained (or understood) if their occurrence relies solely on the presence of an idealization. Initially this seems an intuitive and plausible objection, but if we reflect for a moment on the way that mathematical abstraction is employed it becomes clear that this line of reasoning quickly rules out explanations of the sort we deem acceptable in other contexts. Here the distinction I mentioned at the beginning between idealization and abstraction becomes especially important. Specifically, we need to distinguish between the kind of abstraction that is, in some sense, dictated by our models and the more straightforward kinds of mathematical idealizations that are used simply to facilitate calculation. In the former case the abstraction becomes a fundamental part of how the system is modeled or represented and consequently proves crucial to our understanding of how the system/phenomena behave.

For example, consider the intertheoretical relations that exist in fluid mechanics between Navier–Stokes equations and the Euler equations or between theories like wave optics and ray optics, and classical and quantum mechanics. Because of the mathematical nature of physical theories the relations between them will typically be expressed in terms of the relations between different equations/solutions. In each case we are interested in certain kinds of limiting behavior expressed by a dimensionless parameter δ. In fluid dynamics δ is equal to 1/Re (Reynolds number), and in quantum mechanics it is Planck's constant divided by a typical classical action ($\hbar/S$). But, in fluid mechanics (as in the other cases listed earlier) the limit $\delta \rightarrow 0$ is singular and it is this singularity that is responsible for turbulent flows.

Similarly, in the ray limit where geometrical optics accurately describes the workings of telescopes and cameras the wavelength $\lambda \to 0$. Because ψ is nonanalytic at $\lambda = 0$, the wave function oscillates infinitely rapidly and takes all values between +1 and –1 infinitely often in any finite range of x or t. A good deal of asymptotic behavior that is crucial for describing physical phenomena relies on exactly these kinds of mathematical abstractions. What we classify as "emergent" phenomena in physics, such as the crystalline state, superfluidity, and ferrogmanetism, to name a few, are the results of singularities and their understanding depends on just the kinds of mathematical abstractions already described.

How then should we think about these kinds of mathematical abstractions and their relation to physical phenomena? My point is that the abstract (mathematical) representations supplied by our models are what forms our understanding of these systems. In the case of phase transitions there are formal accounts or definitions that appeal to zeros in the partition function, changes in symmetry and orderliness, and the existence of fixed points.[19] In each of these cases a sharp phase transition is possible—the transition temperature is well defined and the appearance of new orderliness is abrupt. Similarly, these formal features function as indicators of the kind of phenomena we identify with phase transitions—a sharp change in specific volume or density, a change in symmetry or the scaling of properties as measured by critical exponents and correlation functions. In other words, the mathematics provides not only a representation and precise meaning for phase transitions, but it also enables us to associate that representation with dynamical behavior such as symmetry breaking and the appearance of order. The abstract model illustrates the essential features of the phenomenon in question. In that sense the mathematics and the physics are crucially intertwined.

This also has implications for experimental practice. In cases where $N < \infty$ a phase transition is recognized by a finite change in a property like density or magnetization for an infinite change in another property like temperature or the magnetic field (as in the case of permanent magnetization). Yet, in a theory that has only finite volume or finite N we can't be sure that we are identifying a phase transition because the formal continuity of the pressure-volume curve is guaranteed by the analyticity in the activity for finite N. In these cases any discontinuity or unsmoothness is rounded off or smeared. When we do an experiment we are looking for jumps or discontinuities in the data that cannot be smoothed over. Although these jumps and curves are in the phenomena themselves, it is the job of our models (like the Ising model) to tell us the exact form they will take, for example, logarithmic singularity.[20] This is why the appeal to infinitely large systems is crucial; only there will the appropriate kinks and jumps emerge! Similarly in population genetics we need the assumption of infinite populations to determine whether changes in gene frequencies are the result of selection.

If one accepts my characterization of abstraction, then subscribing to Callender's and Earman's account would be tantamount to ignoring large portions of both the mathematical and physical foundations of our theories. On their idealization view we need an answer to why cases like the thermodynamic limit is considered illegitimate when sending distances and time intervals to zero in the use of differential equations is not. Here my distinction between idealization and abstraction offers some help. We typically think of an idealization as resembling, in certain respects, the concrete phenomenon we are trying to understand. We usually know how to correct or compensate for what we have left out of, or idealized in, the description. That is, we know how to add back things like frictional forces that may not have been needed for the problem at hand. Or, we know how to change the laws that govern a physical situation when we introduce more realistic assumptions, as in the move from the ideal gas law to the van der Waals law. These cases are best thought of as idealizations that represent a physical situation in a specific way for a specific purpose.

Constrast this with the kinds of mathematical abstractions described earlier. In the case of the thermodynamic limit, we don't introduce abstractions simply as a way of ignoring what is irrelevant to the problem or as a method for calculational expediency. Instead, the mathematical representation functions as a necessary condition for explaining and hence understanding the phenomena in question.[21] Thinking of abstraction in this way sheds a light on the problem Callender and Earman mention because the problem is very different from the more straightforward cases of idealization. The importance of mathematical abstraction in these contexts requires us to think differently about what constitutes explanation and understanding, but that challenge seems unavoidable. If physical phenomena are and in some cases *must* be described in terms of mathematical abstractions, then it seems reasonable to expect that their explanations be given in similar terms.

We can see how this situation differs significantly from the way fictional models are used to convey information. Although fictional models may constrain the physical possibilities once the model structure is in place, there is typically a choice about how to construct the model and how to represent the system/phenomenon. There is less freedom of movement with idealizations in that specific kinds of approximation techniques inform and determine the way the system is represented. That said, we often have a choice about which parameters we can leave out or idealize, given the problem at hand. In cases of mathematical abstraction, however, we are completely constrained as to how the system is represented, and the abstraction is a necessary feature of our theoretical account.

6. CONCLUSIONS

One of my goals in this chapter was to introduce distinctions among the processes involved in constructing a fictional model, an idealization, and mathematical abstraction. Each of these categories transmits knowledge despite the presence of highly unrealistic assumptions about concrete systems. Indeed, examples of the sort I want to classify as abstractions require these types of assumptions if we are to understand how certain kinds of phenomena behave. Although there is a temptation to categorize any type of unrealistic representation as a "fiction," I have argued that this would be a mistake, primarily because this way of categorizing the use of unrealistic representations tells us very little about the role those representations play in producing knowledge. That said, fictional models can function in a significant way in various stages of theory development. However, in order to uncover the way these models produce information we need to pay particular attention to the specific structure given by the model itself. There is no general method that captures how fictional models function or transmit knowledge in scientific contexts; each will do so in a different way, depending on the nature of the problem. By separating these models from other cases of abstraction and idealization we can recognize what is distinctive about each and in doing so understand how and why unrealistic representations can nevertheless provide concrete information about the physical world.[22]

NOTES

1. For an extended discussion see Morrison (2002).
2. For more discussion of nuclear models see Morrison (1998) and Portides (2000).
3. In other words, is the presence of idealized assumptions different from assumptions that are false in the sense of being deliberately false, like assuming that the ether is made up of rotating vortices of fluid or elastic solid particles? No amount of correction changes the form of these latter assumptions.
4. Although we typically say that models provide us with representations, I think we can also extend that idea to laws. By specifying the conditions under which the law holds (even if they are just straightforward ceteris paribus conditions) we have specified a scenario or a context that defines the boundaries for the operation of the law. This specification can be understood as a representation of the context under which the law can be assumed to hold.
5. Cartwright (1989), in particular, characterizes abstraction as the "taking away" of properties that are part of the system under investigation. For example, when modeling a superconductor one abstracts the type of material the superconductor is made of.
6. As we shall see later, it isn't really the presence of idealizing conditions that is responsible for transmitting information; rather, it is the what is presupposed in the binomial formula for predicting genotype frequencies.
7. All references to Maxwell's papers are contained in the 1965 edition of collected papers.
8. This is what Seigel (1991) refers to as the "standard account." He also remarks (p. 92) that according to the standard account the motivation for introducing the displacement current was to extend Ampere's law to open circuits in a manner consistent with Coulomb's law and the continuity equation. Although

I agree with Seigel that this was certainly not Maxwell's motivation, I think one can incorporate aspects of the standard account into a story that coheres with the historical evidence. My intention here is not to weigh in on this debate but simply to show how various aspects of the field equations emerged from the model.

9. It is perhaps interesting to note here that in the modern version the electric current J appears on the right because it is regarded as the "source" of the magnetic field H. This is because charges and currents are typically seen as the sources or causes of fields. This, however, was introduced by Lorentz, who combined the Maxwellian field theory approach with the continental charge-interaction tradition, resulting in a kind of dualistic theory where charges and currents as well as electric and magnetic fields are all fundamental, with the former being the source of the latter.

10. See Seigel (1991, pp. 130–135).

11. In addition to the agreement between the velocity of propagation of electomagnetic waves and light there were also two other connections between electromagnetic and optical phenomena that emerged from the model; one was the Faraday effect and the other involved refractive indices of dielectric media.

12. In the end, however, there was no real explanation of the way that elastic forces produced electric lines of force. Instead, he was able to calculate the resultant electrical force without any precise specification of how it arose (i.e., there was no calculation of a stress tensor of the medium, as in the magnetic case, from which he could then derive the forces).

13. For an extended discussion of the differences in the two accounts of displacement see pages 146–147 and 150–151 in Seigel (1991).

14. See Morrison (2000) for a detailed account of the development of the electromagnetic theory and the unification of electromagnetism and optics.

15. Why this is an instance of both abstraction and idealization will be discussed later.

16. The law of segregation refers to the fact that the characters that differentiate hybrid forms can be analysed in terms of independent pairs; that is, each analagen acts separately—they do not fuse. We can also understand this as stating that any hybrid for a given character produces an offspring distributed according to definite proportions. If the pure parental forms are A and a and the hybrid Aa, then the offspring of the hybrid will be distributed according to the ration 1A:2Aa:1a. Pearson (1904) was probably the first to show the relation between the law of segregation and the stability of a population in the absence of selection.

17. The notion of presupposed that I have in mind here is the same as the one connected to the ideal gas law—in order for the law to hold one must presuppose that the molecules of the gas are infinitesimal in size and have no forces acting between them.

18. In addition to the presence of unrealistic assumptions about the kinds of populations in which the law holds, the stability embedded in the structural form of the law is also crucial in explaining how/why it works. The variation preserving feature associated with the stability requires that the gene have a discrete, atomistic character. The importance of the "unrealistic" constraints is to highlight the impact of different factors and the degree to which they affect the evolutionary processs. Because the law represents a stable stationary state for a sexually reproducing population, it enables us to judge the effects of selection, preferential mating, etc. on the homogeneous field of allele and genotype frequencies. The *fundamental assumption* required for this picture to work is that of a structure that preserves modifications.

19. The most important information in the renormalization group flow is given by its fixed points, which give the possible macroscopic states of the system at a large scale.
20. Similarly, the uniformity of convergence defines what we mean phenomenologically by a phase of matter.
21. For an excellent discussion of issues surrounding the notion of reduction and the thermodynamic limit see Batterman (2005).
22. Support of research by the Social Sciences and Humanities Research Council of Canada is gratefully acknowledged.

Part IV
Fictions in the Physical Sciences

8 When Does a Scientific Theory Describe Reality?

Carsten Held

1. INTRODUCTION

Scientists often claim explicitly that their theories describe portions or aspects of the real world, but is this self-assessment really correct? Do scientific theories describe the world? Do they aim at such description? Should they do so? Philosophers of science, in contrast with its practitioners, are notoriously divided over these questions and the divide is one demarcation in the scientific realism debate. Scientific realists fully accept the sciences' self-appraisal and hold that theories indeed aim at accurately describing the world and progressively approximate this aim. By contrast, for their antirealist opponents an epistemologically legitimate goal of science is just to devise structures generating acceptable explanations and approximately true predictions for those phenomena we directly observe.

So far, so well known. However, the philosophical dispute over scientific realism usually translates the question whether theories describe into the one whether all suitable expressions of a theory are meant to refer, and in an established theory typically do refer, to real-world entities. This translation seems to rest on two misconceived presumptions about what happens in real science—one epistemological, one linguistic. The epistemological presumption is that scientists (and educated laypeople alike) assume theories from the start to aim at describing the world, assume their expressions from the start to aim at referring to real-world entities, and assume good (mature, well-confirmed) theories to more or less achieve these aims. The epistemological debate, accordingly, is taken to concern only whether this assumption is justified, whether the philosopher can go along with the naive mind or must practice epistemological modesty. Just as we generally assume that we possess instances of knowledge, but epistemologists quarrel about our justification in doing so, philosophers of science ask whether scientists are justified in assuming that all their propositions and expressions refer to the world—presupposing that scientists do assume this.

The second presumption in the debate is that the question of reference concerns linguistic expressions (perhaps sentences, but mostly, terms) as types, not tokens. The question about the reference of theory expressions, from Carnap's and Grover Maxwell's time on, concerned terms like, say, 'electron,' regardless of the context of a scientific theory wherein tokens of 'electron' appear. Especially, this discussion has been insufficiently sensitive to whether the expression 'electron' appears in the presentation of a theoretical model or in the context of identifying the cause of a naked-eye observation, in an undergraduate textbook or the description of a new experiment in a learned journal. But this second presumption means that the question of what an expression like 'electron' refers to is one that can be posed quite generally, out of any context of use, and that scientists themselves tacitly assume it answered in this general way. If scientists naively believe that their theories describe, say, electrons, then they seem to believe that any occurrence of 'electron' anywhere in a theory refers to real electrons.

Taking both presumptions together, we get an implausibly static picture of real-world scientific theories. Expressions taken by scientists to refer to real-world entities at the point when a certain part of some theory is indeed taken to describe reality are assumed as potentially referential throughout the theory, because it is viewed as a description of reality, *tout court*. But this is an unreasonable belief. As counterevidence, take a theory plus some assumptions specifying a model such that it instantiates the theory. Take the model to be an abstract structure with ideal properties such that the theory's sentences, equations say, apply to the structure in a particularly simple manner. The expressions used to describe the model cannot describe reality, nor are they intended to do so—as witnessed by the ideal properties. For example, in the description of a model where an electron is assumed to be a point mass, the term 'electron' cannot reasonably be understood to refer to real electrons. Now, assume that the theory's vocabulary were homogeneous with respect to different tokens of an expression. Then, if some other part of the same theory is conjoined with different assumptions and is taken to describe reality, its stock of expressions cannot include the same expressions as before, for example, not the term 'electron.' The theory would have to be divided into two compartments with different vocabularies—which for real-life theories obviously is not the case.

We thus have good reason to believe that token expressions of the same type often have quite different functions in different parts of a theory. Theories are differentiated and inhomogeneous entities with respect to their different parts' references to the world. Within the concrete presentation of a theory this fact materializes in the different referential functions of token expressions. So, before we contemplate the question: Can a scientific theory legitimately be taken to describe reality?, we had better answer the one: Does it aim at such describing? It would be silly not to answer this second question in the affirmative *in some sense*. However, as I have suggested, it also can be asked meaningfully when, or at which stage, a theory aims at describing reality. And it is this question I want to consider here. Assuming that something, if intended to describe, describes either truly or falsely, so describes in any case, I can put my question in a slightly slimmer form: *When does a scientific theory describe reality?*

I have suggested an internal differentiation among a theory's token expressions as to their referential function. This differentiation can be grasped quickly by looking into the different functions one and the same theory can have. Theories, after all, serve many purposes, and some of their functions are more basic and less controversial than their eventually being world descriptions. In particular, a scientific theory can aptly be characterized as a tool for explaining and predicting phenomena within a certain domain. Theories first and foremost are generators of hypotheses participating in the making of predictions and potential explanations, but these hypotheses do not from the start aspire to describe anything real. When a hypothesis generated from some theory participates in the successful explanation or prediction of phenomena, this success corroborates the theory. In case of recurrence, the latter can then be elevated to the rank of a description of reality. This will be the case when an explanation, because of its success, leads us to infer the entities and properties it postulates or when a prediction's success is counted as evidence for the truth of its generating premises. Hence, bits of theory have different functions at different stages of theory acceptance, and the same will be true for a theory's concrete linguistic representation. The expressions used to present a theory will have different referential functions depending on their loci not within the theory, but within the theory's *presentation*. Does the theory's presenter signal that it is now used to generate a prediction or a potential explanation? It will be unlikely, in this context, that an expression is used to describe reality. Does the presenter signal that the whole theory, because of its eminent explanatory or predictive success, can be viewed as describing reality? Then an expression is likely to be used descriptively. And because the presenter can signal a change of attitude toward a formerly putative explanation, it may even be the same token expression that changes its role in retrospect. We must view the relation of linguistic theory presentations and their references as differentiated with respect to the different contexts of tokens of an expression, and even as dynamical with respect to later reinterpretations of these token contexts.

All this is best illustrated by an example. In what follows I will take a detailed look at a concrete presentation of a piece of physical theory (excerpts from a treatise on quantum field theory, henceforth briefly QFT) to analyze the referential relations. The example chosen signals an explanatory function, and I will ignore the complication that the presentation might alternatively be viewed as the prediction, rather than the explanation, of an effect.[1] The explanandum effect is the Lamb shift in excited hydrogen atoms, and because QFT explains it with extraordinary precision the theory usually is taken to describe simpliciter a realm of microentities producing this effect, among others. Such a general descriptive attitude toward QFT could be evidenced from an undergraduate textbook that just reports the state of the art about the Lamb shift in light atoms, without aiming for a detailed explanation. Interestingly, this attitude is missing in the text at hand, which does report a detailed explanation of the shift. Within this explanation a different attitude is present where single bits of theory, that is, token expressions, are clearly not intended to describe anything real. And although we may surmise an overarching intention to describe reality from our contextual knowledge of QFT's power and explanatory success, there are very clear and explicit linguistic signals for the opposite attitude in the details of the presentation. Its author signals in nearly every step that his declaratory sentences do not aim to describe reality. The whole presentation thus confirms that it is during its application for an explanatory purpose—indeed, only at the very end of the explanation—that the theory makes contact with the real world.

The Lamb shift case is just an example and cannot replace a general argument. But I take this example to be typical not just for physical science, but for science in general. I take the explanatory example to have a counterpart in clear predictive applications of a theory. And I take it that theory parts constitutionally preceding a theory's application have even less contact with reality than this application has. These three theses, however, are not defended here. Instead, I concentrate on my examples and show how a typical scientific theory makes contact with reality.

A recent contribution to the scientific realism debate invites its readers to "crack open a [science] textbook" of their choice and predicts what they will find: "No matter the field, at the heart of the text you will typically find a theory *purporting to describe* some part or aspect of the world" (Stanford, 2006, p. 3; my emphasis). This is indeed true for those theories we pack into textbooks and teach our science major undergraduates, that is, for established theories, to which we take the attitude of *accepting* them as world descriptions because of their predictive and explanatory success. But it is true only given this attitude. Toward a theory that is presented in order to evidence its predictive and explanatory merits, we take a very different attitude. In case of such a presentation, neither presenter nor audience takes the theory from the start as describing reality or purporting to do so.

2. EXPLAINING THE LAMB SHIFT IN QFT

In his 1995 book on the foundations of QFT, Steven Weinberg demonstrates how the theory explains the Lamb shift, a splitting of energy levels in hydrogen atoms that was not predicted in its precursors (most importantly, Dirac's 1928 theory of the electron), but was found by Lamb and Retherford in experiments published in 1947. This is an example of scientific explanation from an established and well-confirmed theory. Weinberg's book, though featured as a textbook for graduate courses in theoretical physics, is not elementary. In the case of the Lamb shift, the text conveniently sums up how one of our most successful theories explains a crucial effect in its domain. It thus provides a first-rate example of scientific explanation where we can study how a theory is applied and thereby makes contact with reality. To do so we will have to look into some of the linguistic details of Weinberg's presentation.

In a historical chapter, Weinberg describes Lamb and Retherford's original experiments and physicists' attempts to come to terms with them (1995, sec. 1.3, pp. 31–36). Much later (in sec. 14.3 'The Lamb Shift in Light Atoms,' pp. 578–594), the effect is derived from a fully developed QFT. The reader can view the derivation as an explanation of the effect, reported earlier, from the theory. (The numerical result derived is, in fact, confronted with experimental results from the 1980s, advancing the exactness of Lamb and Retherford's experimental value [circa 1000 MHz] for the shift very much. The derivation may be seen as an explanation of these results, too, or as a prediction, tested by them; Weinberg does not make this fully explicit.) The explanation itself comprises 17 pages of hefty calculations. I quote and analyze just those parts of the description of the explanandum (i) and the explanation proper (ii)–(iv) that contain interesting problems of the reference of token expressions.

2.1. Reporting the Explanandum

Here is an excerpt from Weinberg's account of Lamb and Retherford's experimental results, in Chapter 1 of his book:

(i) . . . in the late 1930s . . . spectroscopic experiments began to indicate the presence of a $2s_{1/2}$–$p_{1/2}$ splitting of order 1000 MHz . . . [At a conference in 1947, experimentalist] Willis Lamb described a decisive measurement of the $2s_{1/2}$–$p_{1/2}$ shift in hydrogen. A beam of hydrogen atoms from an oven, many in 2s and 2p states, was aimed at a detector sensitive only to atoms in excited states. The atoms in 2p states can decay very rapidly to the 1s ground state by one-photon (Lyman α) emission, while the 2s states decay only very slowly by two-photon emission, so in effect the detector was measuring the number of atoms in the metastable 2s state . . . [Finally,] the intrinsic [i.e. $2s_{1/2}$–$p_{1/2}$] splitting could be inferred. A preliminary value of 1000 MHz was announced, in agreement with the earlier spectroscopic measurements. (Weinberg, 1995, pp. 34–35)

From the second sentence in (i) onward, Weinberg reports Lamb's own description of his and Retherford's experiment. He thus refers to a single experiment on a certain beam of hydrogen atoms generated in a laboratory at Columbia University in 1947. His explanation of the effect more than 500 pages later does not start to reestablish reference to these hydrogen atoms—and no one would expect it to do. Of course, Lamb described an experiment considered to be repeatable and thus he reported an effect considered to be observable in all hydrogen atoms in the world. So, the behavior of certain atoms in an exactly described and repeatable experimental situation is inductive evidence that all hydrogen atoms in this situation would behave similarly and exhibit the energy difference now called Lamb shift. Accordingly, Weinberg freely switches between referring to the hydrogen atoms in Lamb and Retherford's experiment and the set of all real hydrogen atoms. Reference to the set of all entities of a kind is generally signaled by switching to the atemporal present tense—and Weinberg also employs this linguistic device. In the opening sentence, the nominalization "the presence" could be resolved into the sentence 'there is a $2s_{1/2}$–$p_{1/2}$ splitting of order 1000 MHz,' thus explicating the atemporal present. In the fourth sentence, Weinberg explicitly uses a verb ("[can] decay") in atemporal present to describe general properties of hydrogen atoms, but then returns to the concrete atoms of the 1947 experiment.

This linguistic evidence is significant. I want to show that QFT, in the course of being applied within an explanation, makes contact with reality by making contact with the explanandum. But the explanandum is not a report of raw data and thus is not itself an unmediated description of empirical facts. It is important therefore to fix the sense in which the explanandum nevertheless *is* a description of reality. Let us consider the following sentence, inferred from the Weinberg text, as the explicated explanandum: 'In excited hydrogen atoms there is a splitting of $2s_{1/2}$–$p_{1/2}$ levels of order of 1000 MHz.' Having thus a concrete explanandum sentence before our eyes, we can quickly clarify the referential situation. In this sentence, the atemporal present tense of the verb signals a generic meaning, which makes the whole sentence acquire generic meaning.[2] In particular, also the noun phrases have generic reference; that is, the phrase 'excited hydrogen atoms' refers to the collection of all members of the class of relevant hydrogen atoms. In such a case, the intention clearly is to transcend the set of actually observed objects, that is, the excited hydrogen atoms observed by Lamb and Retherford, but the intended class is not the one of all real excited hydrogen atoms. Indeed, it is characteristic for generic reference that an individual in certain exceptional circumstances does not count as a counterexample, and thus is not assumed to be a member of the class. An excited hydrogen atom that due to an ingenuous experiment was prevented from decaying into the ground state would not count as falsifying the generic statement. The latter is naturally read as referring to atoms in appropriate circumstances only. The generic reference is, in this sense, robust with respect to outliers.

It is very plausible to consider our constructed generic statement, rather than a report about concrete cases, as an explication of Weinberg's explanandum. When we ask whether the explanation of the Lamb shift from QFT is aimed at the effect reported by Lamb and Retherford (or, for that matter, the set of all Lamb shift measurements done since 1947) or at the Lamb shift, *simpliciter*, the latter clearly is the appropriate option.

We expect that a scientific explanation, except in special circumstances, covers the generic case rather than just a finite number of instances. So the explanandum to consider is the generic, which clearly transcends the actually observed data. However, despite transcending the data the generic statement is equally clearly meant to be a description of reality. Here it ascribes to all real hydrogen atoms a dispositional property they will actualize in appropriate circumstances.[3]

Thus, brief reflection confirms what we would have presupposed from the start, that is, that our explanandum sentence describes a real phenomenon. The referential situation is more complex and mediated than one would have thought, but the descriptive attitude is in place. What I want to point out is something else. The specific linguistic means employed to communicate this attitude can be put to an entirely different use in another context. Indeed, the generic reference, intended here to refer to objects in the real world, can be expressed in a way that in isolation is indistinguishable from a situation where such reference is not intended. To see this we have to look at yet another variant of our explanandum. Initially, consider definite singular noun phrases like 'the hydrogen atom.' Generic reference is largely insensitive to the use of singular or plural, definite and indefinite article. At least 'hydrogen atoms,' 'a hydrogen atom,' and 'the hydrogen atom' can express such reference (but 'the hydrogen atoms' would be unusual). I take it that in such a case 'a hydrogen atom' refers to an arbitrary member of the class of such atoms and 'hydrogen atoms' to an arbitrary plural of members. What about 'the hydrogen atom'? Indeed, singular definite noun phrases can express generic statements, but is this phrase really referring to all class members? How then is the singular explained? Without arguing the case in detail here, I propose that this phrase does refer to all members of the class (a plural), but indirectly, that is, mediated by a *representative* of the class (a singular). Thus consider the following variant of our above explanandum: 'In the excited hydrogen atom there is a splitting of $2s_{1/2}$–$p_{1/2}$ levels of order 1000 MHz.' To have a handy name, let's call such a generic sentence with a singular definite noun phrase a *singular generic*. Clearly, this singular generic is meant to report the same fact as before. It transcends the concrete observations of (let's say again) Lamb and Retherford, but is meant to cover the concrete cases insofar they are relevant. In this sense, the sentence describes reality. But it does not do so directly. Reference is mediated by an individual: *the* typical or otherwise relevant 'excited hydrogen atom,' representing all members of the class of such atoms. But the entity representing the class members does not need to, and in general is not, a class member. The typical entity, in this or that respect, is in general, an abstract entity. In the case of a singular generic, it always is. Thus, the subject noun phrase in our singular generic effects its reference to real entities via an abstract one.

Generic sentences transcend the observed, but aim at describing reality and so does the singular generic. What we have, after brief analysis, is a quite complex case where the intention is to describe real objects, but reference to them is mediated by an abstract entity. Now, what does 'mediation' mean here? Does it mean that the phrase *refers* to the abstract entity—and then refers further to the class members—or does that entity mediate reference without being itself referred to? It seems difficult to conceive of the mechanism of reference in a way allowing that an expression can refer to different entities (here: a singular and a plural of entities) at once. So the first option is implausible. Instead, the mediating entity should not count as being referred to in the mediation process. But this concerns the expression in the mediating process, that is, its generic use only. Consider that, vis-à-vis our sample generic containing 'the excited hydrogen atom,' an interlocutor asks, 'Which hydrogen atom?' This would, of course, be an ill-posed question betraying a misunderstanding of the author's intention. The author, by means of the singular noun phrase, means to refer to all class members, but does not mean to refer either to one of the members, or to the abstract proxy used to represent all class members. The abstract entity used is out of focus of the understanding audience and comes to the fore only given the imagined ignorant question. The author may counter it by rephrasing and thus decoding the original singular phrase: By 'the hydrogen atom' I mean all hydrogen atoms in suitable circumstances. The author may as well counter the question by establishing a reference that was not there before: By 'the hydrogen atom' I mean an abstract representative of all members of the class; it makes no sense to request this abstract entity to be pointed out. It is only in the last case that an abstract entity has been referred to.

A singular noun phrase like 'the excited hydrogen atom' is crucially context-sensitive. Indeed, although here the phrase is meant to refer to all members of a class of real objects by means of an abstract proxy that is not itself referred to, in other contexts a similar phrase does refer to an abstract entity, and one that is not a proxy for anything. (An illustration will immediately follow. *Vide* all noun phrases discussed in sec. 2.2.) For now the fact that, in the singular generic, reference is mediated by an abstract entity allows making contact with a well-established conception in the theory of scientific theories: the *data model*. Indeed, the data model is a plausible candidate for the referent of a singular generic noun phrase. Consider what has to be done when constructing a data model from raw data. The first step is to eliminate outliers from the data set. They are considered the result of erroneous observation or otherwise untypical for the set. The second step is to extrapolate from the data to what could have been, but was not, observed.[4] Both these activities are clearly present when observations of single cases are represented in a generic sentence. Observations we consider erroneous are always excluded when we generalize from what we think we have correctly observed. But when we report our generalization in a generic sentence this will express our neglect of other outliers as well—because the generic sentence is insensitive in meaning to the untypical cases. Moreover, such a sentence expresses an extrapolation beyond the data. Finally and most importantly, the singular generic can be read such that there is direct reference to an abstract entity, but this entity itself represents the single cases. On the other hand, every data model worthy of its name models something, i.e. is a representation of the data. So, in a straightforward sense the abstract entity mediating reference to real entities in a singular generic can be identified with a data model. However, it will be suitable for the ensuing discussion to again simplify matters and speak of reference to a *phenomenon* here. This expression glosses over the difference between the mediating abstract entity representing the real ones in the singular generic (and sometimes being referred to) and reference to the real ones *simpliciter* in other generic sentences, but it is most natural and highlights the common intention, in all these cases, to describe an aspect of reality. Moreover, the usage tallies with the theory of Bogen and Woodward (1988), who interpret phenomena as regularities extrapolated from the raw data.

All in all, a generic sentence based on information from single observations expresses the intention to refer to entities in the real world, and even the mediation, in a singular generic, by an abstract proxy does not foil this intention. An educated audience will simply read the phrase 'the excited hydrogen atom' in the Weinberg explanandum as referring to all members of the class of relevant excited hydrogen atoms in the real world, even if on reflection the linguistic mechanism functions via a singular entity we cannot localize out there in the world. Assuming for a moment that the representing singular entity in some sense exists, we can use a well-known philosophical metaphor to describe the referential situation: The singular entity mediating the reference is *transparent* in the sense that the expositor does not mean to address it, but through it intends to address real-world entities. The key point here is this: In another context a token of the same singular definite noun phrase can appear without any intention to describe reality. This is the case I will examine next.

2.2 Introducing a Theoretical Entity

More than 500 pages later, in Chapter 14 of his book, Weinberg treats the Lamb shift theoretically. Section 14.3 starts thus:

(ii) Let us now consider radiative corrections to the energy levels of a nonrelativistic electron in a general electrostatic field, such as an electron in the Coulomb field of a light nucleus . . . It is natural in this limit to treat the Coulomb field as a weak perturbation. (p. 578)

The reasoning starting here will ultimately discharge an explanation of the Lamb shift in hydrogen atoms. Initially, however, it makes no reference to such atoms and will not do so until the very end (quoted in (iii) later). This is no accident. Weinberg emphasizes the wider scope of his account by his more general section title: "The Lamb Shift in Light Atoms." Hence, the explanation we are interested in is the result of theoretical considerations aspiring to cover a broader range of phenomena. To achieve this, Weinberg starts the section (and thus the material that will later be integrated into the explanation) by inviting the reader to consider an entity that could be viewed as the crucial constituent part of a hydrogen atom or another light atom: an electron. Thus, he asks the reader to focus on "a non-relativistic electron" in a specific situation. (Literally, the entity considered is a property of such an electron, but let's ignore that complication.) What is happening here? The opening sentence is an imperative, a request for the reader to consider an entity of a certain kind. Two of its components are crucial here. First, the indefinite noun phrase signals the introduction of a new entity into the discourse. Speech acts containing such phrases direct an audience's attention to an entity for future reference. This linguistic phenomenon is well researched in pragmatics and there are criteria for deciding whether an entity so introduced is entirely new to the audience or not.[5] Trivially, new entities are introduced into a discourse not only in scientific contexts, but in all kinds of communication, especially also in descriptions of real-world situations. The linguistic means are similar in all cases, and the use of an indefinite noun phrase like "a non-relativistic electron" is only one of them. Second, we have the introduction of a new entity within a *request*, something we seldom encounter in everyday contexts. Moreover, this never happens in a world describing sentence and again trivially so, as descriptions are not requests. The request we have here to "consider" an entity invites the audience to imagine or otherwise grasp an entity. We can leave open here whether this is a genuine contact with an entity of a special kind, that is, whether this entity in some sense exists. What matters is that this request introduces the audience to a new entity, but nevertheless does not establish any referential contact with a real object.

So, quite unsurprisingly, Weinberg's reasoning in (ii) does not start by establishing reference to a real object. An indefinite noun phrase alone could be understood as referring to some unspecified worldly object, but this phrase would have to be part of a descriptive sentence, not a request. (Pertinent examples of such phrases are "a beam," "an oven," "a detector," in (i) shown earlier.) One might wonder whether Weinberg's reasoning could not be interpreted as addressing a phenomenon in the sense explained earlier. After all, the phenomenon in one sense is an abstract entity and thus might be available here. If this were plausible, Weinberg would be addressing an abstract entity directly, but real ones indirectly, and there would be the intention of some referential contact with reality. But this is obviously implausible. The opening of a theoretical reflection is not the point where reference to a phenomenon is in place. More technically, any establishing of a referential contact with reality, however mediated, would require a descriptive sentence, not a request.

I shall call a request like Weinberg's a *theoretical request* and the entity it introduces a *theoretical object*. Now, the opening sentence of (ii) does not only introduce a theoretical object, but does so as part of an exposition that later on functions as an explanation. This exposition deduces, in the object introduced, an effect that parallels the explanandum and when the parallel is drawn the explanation is completed. (This point will be reached, by way of an allusion only, when Weinberg speaks of an "agreement" of values; see (iv) in 2.4 later.) I shall say that, within a scientific explanation, the deductive or expository part describes a theoretical entity and thereby exposes what ultimately becomes a *theoretical model*.

These expressions are surely not innocent and reveal a theoretical commitment. I speak of theoretical request and object because the one introduces the other in the typical course of developing a scientific theory. I presuppose here that Weinberg's exposition is a core piece of QFT proper. Indeed, I think that expositions like this, blending over into explanations (which clearly are applications), cannot be cleanly separated from the theory itself as mere applications. However, for those who think that such a separation makes sense one might point out that similar requests to consider objects appear also in those parts of a theory that are not considered applications. I do not take a stand here on the idea of theoretical entities as those entities that, in the discussion about scientific realism, the realist takes a theory to postulate as existing. Obviously, the phenomena are different. A theoretical object in the present sense is interpreted as something that the author does not necessarily take to exist. By contrast, a theoretical entity in the scientific realism sense is one that a philosopher of science of realist persuasion takes to exist because, allegedly, our present successful theories tell us that it exists while the antirealist finds that our theories do no such thing and counsels agnosticism. Both conceptions are certainly related, but further discussion transcends the scope of my present project.

Now, I also say that describing a theoretical object, in the context of a theory's application for explanation, amounts to introducing a theoretical *model*. Here, I mean to describe the following. In the exposition of a theory a theoretical entity may be introduced. This initial step limits the further course of exposition in no way. If the theory is developed, like in the present case, to ultimately discharge an explanation, then the final step establishes an explicit contact with reality via comparison with an explanandum phenomenon. If we call something a scientific or theoretic model, we usually employ the intuitive (nonlogical) sense of 'model.' In this case, we are committed to assuming that there is a target entity that is intended to be modeled. 'Model' in this sense means a kind of representation, which concept entails that there is the intention to represent an entity, the model's representandum. If one thinks that scientific explanation operates the way I have outlined, that is, that in a theoretical 'story' or in an account of a theoretical object an effect is demonstrated or exhibited that parallels the explanandum phenomenon, and if one further thinks that this object on which demonstration or exposition takes place is a theoretical model, then there is at least a part of the model, that is, the effect parallelizing the explanandum that ultimately does model, that is, represent something real. So, in short, a theoretical model functioning in the way just sketched within an explanation naturally exhibits a contact with the real world—at least in the sense of modeling the explanandum phenomenon.

This is, again, a simple observation. But it adds to the complexity of the referential situation. In particular, one might get an impression of outright inconsistency that needs to be dispelled. I claim that the theoretical object introduced in a theoretical request is neither real nor in any intended representational contact with a real entity. But I also claim that such an object, in the course of an explanation, is given the role of a theoretical model. A model, by stipulation, is meant to represent something real: the model's target entity. Thus, now the same object I addressed before is taken to be in representational contact with a real one. But there is really no inconsistency here if we understand the situation dynamically and distinguish a context where an expression referring to a theoretical object serves to introduce and then further describe the object from a context where the expression is used to signal that the object now represents something in the real world. In the first context, the expression does nothing but introduce the object and help describing it. There is no reference to anything real here. Only when a feature of the entities thus described is finally used within an explanation, the whole construction is promoted to the status of a theoretical model. From this point on, the entity in question is intended to represent something real, and eventual expressions referring to the model entity now have an indirect contact with reality. This is to say that there is a certain latest point in the exposition, from which on the modeling purpose becomes visible. Unsurprisingly, real-life cases are not as clear-cut and explicit, in this respect, as the philosopher of science would wish for. In Weinberg, we will find only expressions referring to theoretical entities *simpliciter*, but none referring to theoretical entities turned models. Nevertheless, there is a latest point also in this example from which on the modeling purpose is understood clearly.[6]

By means of the Weinberg example, I have described the construction of a theoretical object that, in view of its final function, can be characterized as a would-be model of a phenomenon. In the process of communicating this construction, singular noun phrases have a function entirely different from the one analyzed before (in sec. 2.2). Such noun phrases, whether indefinite ("a non-relativistic electron") or definite ("the Coulomb field"), are meant to introduce and help describing a theoretical object. Especially, a phrase like "the excited hydrogen atom" as in our previous reconstruction of the explanandum might appear as well. But whereas in the explanandum the reference to real-world objects or phenomena could easily be made explicit, here no such reference is intended and hence cannot be so explicated. Only when, in the end, the theoretical entity in the end is viewed as a model it acquires a representational capacity. Expressions exposing the model refer to the model entities, which in turn are meant to represent worldly phenomena. The referential situation thus is still markedly different from our preceding description of a phenomenon. In the metaphoric language introduced earlier, the theoretical entity introduced by Weinberg is *opaque*, rather than transparent, because it is a representation of something else, but the expositor's expressions mean to address the representation, rather than its representandum.

2.3 Exhibiting an Effect in the Theoretical Entity

Weinberg's reasoning, preparing his explanation of the Lamb shift, started out with a theoretical request introducing a theoretical entity. This entity is now stepwise attributed well-defined properties allowing a mathematical treatment. Weinberg aims at ultimately constructing an object with the key properties of a real hydrogen atom, one that can be viewed as a model of real hydrogen atoms—which for our purpose we can just identify as the considered electron, not in a general electrostatic potential, but in a Coulomb potential. The hydrogen atom constructed in this process is, of course, a theoretical entity, just like the original electron. For the understanding reader, the explanatory purpose is always in the background, provides the blueprint for constructing the entity, and thus points forward to the later contact with reality. Nonetheless, the constructed entity itself before that contact is ontologically on a par with our theoretical electron. That is to say, referentially it is entirely detached from the real world.

Weinberg's argument is meant to discharge a scientific explanation of the Lamb shift, and if my presumption is correct that such explanation utilizes a theoretical model, which itself is a theoretical entity, then it is this representational capacity of the theoretical entity exhibited that the audience must understand. Perhaps such understanding is in the background all through the exposition. However, it must be present only at the point when the audience is to understand the entity's final role in the explanation. Understanding the representational capacity is not essential to understanding that a theoretical entity is introduced and some of its properties are specified. The entity may be introduced as a potential model, but is not intrinsically predisposed to be one.

I assume that the purpose to use a theoretical entity as a model, within an explanation, becomes finally explicit when it is conceptualized in the way we conceptualize the explanandum. So I assume here (a) that author and audience presuppose hydrogen atoms to be real and to exhibit a certain energy shift between the 2s and $2p_{1/2}$ states in suitable contexts as our explanandum sentence had it and (b) that they wish to explain this very feature. So when Weinberg starts to call the entity constructed from the electron "the hydrogen atom" and explicitly says or otherwise clarifies that it exhibits a certain energy shift between the levels in question, the modeling purpose becomes overt. According to this interpretation, Weinberg at this point is no longer speaking just about a theoretical entity, but about a theoretical model. There now is an indirect referential contact with reality via the modeling purpose. But Weinberg, when using a definite noun phrase like "the hydrogen atom" in this context, is not making reference to anything real, although the same phrase, when used in a singular generic in our explanandum or in a concrete observation report, is used for just this purpose. This interpretation is corroborated directly by looking at the text.

Weinberg (1995), after having given an expression for the energy shift in a general electrostatic field and after having specified for the Coulomb potential, arrives at a numerical result for the Lamb shift. He sketches how theoreticians have improved the calculation by including higher order terms and other effects, mentions the improved results, and finds them in excellent agreement with up-to-date (1981 and 1986) measurements of the shift. This final comparison concludes the explanation. For discussion, we need to have all this material before our eyes in bulk. Here are the two relevant passages[7]:

(iii) The classic Lamb shift is the energy difference between the 2s and $2p_{1/2}$ states of the hydrogen atom, states that would be degenerate in the absence of radiative corrections. Our calculation has given . . . 1052.19 MHz. (p. 593)

(iv) The calculation of the Lamb shift described here has been improved by the inclusion of higher-order radiative corrections and nuclear size and recoil effects. At present the greatest uncertainty is due to a doubt about the correct value of the rms charge radius r_p of the proton. For $r_p = 0.862 \times 10^{-13}$ cm or $r_p = 0.805 \times 10^{-13}$ cm, one calculation gives a Lamb shift of either 1057.87 MHz or 1057.85 MHz, while another gives either 1057.883 MHz or 1057.865 MHz. Given the uncertainty in proton radius, the agreement is excellent with the present experimental value, 1057.845(9) MHz. (pp. 593–594)

In (iii), Weinberg no longer talks about an electron in an unspecified situation, about functions, cutoffs, and the treatment of integrals. Instead, he now addresses his target object, the hydrogen atom, and the Lamb shift. In fact, it is the first place in the section where he mentions the hydrogen atom and the second one where he mentions the Lamb shift.[8] This is further evidence that during the largest part of the argument no contact whatsoever is sought with the explanandum. It is only now, at the end of the explanation, that such contact is established. Isn't it plausible to say that Weinberg now starts to address objects we assume to really exist? Yes, it is indeed, but reference to reality is not established when "the hydrogen atom" is addressed at the beginning of (iii), but only in the final sentence of (iv).

In the beginning of (iii), the phrase "the hydrogen atom" refers to a theoretical entity, which now is obviously meant to be a model of real hydrogen atoms. It is this point I had in mind earlier as the one where the modeling purpose becomes finally obvious. Calling the constructed entity "the hydrogen atom" explicates what it was constructed for and allows for a comparison with real phenomena. Now, in order for the referent of the phrase "the hydrogen atom" to be a model of real hydrogen atoms it must be the case that the phrase is not referring to anything real in the first place. But isn't it instead plausible to say that now (in (iii)!) Weinberg finally is referring to real hydrogen atoms? Perhaps he does so again by means of a generic sentence? My claim is that, despite first impressions, this is not the case, but this must now be substantiated.

The first sentence in (iii) can be interpreted in two very different ways. I will discuss both interpretations, but discard the first in favor of the second. First, "the hydrogen atom" can be interpreted as referring to real hydrogen atoms along the lines of my analysis in section 2.1, earlier. Second, it can be viewed as referring to a theoretical entity. Let's consider both options in turn. The opening sentence of (iii) might be seen as picking up, qualitatively, on (i), the report of the explanandum phenomenon. If this is true, then "the hydrogen atom" here does refer to a phenomenon in the sense explained earlier. However, it seems obvious that the second sentence, running "Our calculation has given . . . 1052.19 MHz," means 'Our calculation of the shift between the states in the hydrogen atom'; that is, the calculation refers to the object referenced just before. So it is the object of his own previous calculations that Weinberg is talking about in both these sentences (in the second one only implicitly). I have indicated that Weinberg, from his original electron, builds up an electron in the right potential with the right excitation energies, that is, something that 'looks like' a real hydrogen atom in the relevant aspects. I have also argued that it is implausible to interpret the original object as a phenomenon. Likewise, the hydrogen atom constructed from the original object is no such phenomenon. This interpretation can be further sustained now. Assume (for *reductio*) that Weinberg is referring to a phenomenon. The reference reestablished at the beginning of (iii) is implicitly kept throughout (iii) and (iv), where (iv) is a paragraph briefly describing improved calculations of the Lamb shift in the hydrogen atom. The capstone of the paragraph (and indeed Weinberg's section, chapter, and the whole volume) is the "excellent agreement with the present experimental value" (final sentence of (iv)). So we must answer, given the present assumption, what Weinberg finally compares and finds to agree with what. Assume plausibly that by "the present experimental value, 1057.845(9) MHz" he means a property shared by all concrete hydrogen atoms (or all those subjected to experiment, it doesn't matter). He would then be comparing a property of the phenomenon (which, by the lights of 2.1, is an abstraction!) with one shared by all of its instances. But the phenomenon just has the properties we learn suitable ('good') single atoms to have. What properties it has is a matter of stipulation and cannot be a merit of resourceful theorizing like the one Weinberg describes. Moreover, if all hydrogen atoms have a relevant property, so should the phenomenon constructed from them. In this case, the relevant property shared by the instances would, by assumption, be possessed by the phenomenon. And this, of course, flies in the face of the textual evidence that Weinberg is comparing *different* numbers. So this is an implausible interpretation. Take it now, again most plausibly, that Weinberg when talking about "the present experimental value" means the value found in present experiments in the hydrogen atom, that is, means the second expression to be the property of a phenomenon. If this is true, we would have to hold that Weinberg is comparing the phenomenon abstracted from real hydrogen atoms and their properties with itself or, if there is more than one such phenomenon, one phenomenon with another. The first possibility is incoherent because of different numerical properties ascribed to the same entity. The second possibility would leave open the question of why we take two phenomena, abstracted from concrete observations, to contain different properties. We take these properties to be shared by all, or many or typical or relevant, individuals, so this result would require explanation. And, of course, we know that the theoretical value is *theoretical*. It hasn't been learned via abstraction from concrete atoms. But now we have exhausted all the options of understanding "the hydrogen atom" in (iii) as referring to a phenomenon, that is, to something real. Thus, the whole interpretation should be abandoned and with it the assumption on which it was based, that is, that there is some kind of reference, however indirect, to a real object in the first sentence of (iii).

What we have, quite trivially, is the comparison of (let's say, for simplicity) two numbers, one derived from a theory, one found experimentally. But these numbers are not numbers without any meaning. They designate numerical properties of entities, indeed of a hydrogen atom that is a theoretical entity and all the real hydrogen atoms, the latter only mentioned indirectly in the phrase "the present experimental value." What we have is, very remarkably, a comparison between theoretical and real entities.

2.4 Establishing Contact With Reality

It is only in this very last step that the whole reasoning makes contact with reality, that is, when Weinberg says at the end of (iv) that "the agreement is excellent with the present experimental value." This experimental value doubtlessly is something found in real hydrogen atoms. It is immaterial here how we construe the referential connection of "the experimental value" with real atoms. Regardless of whether the construction requires mediation by an abstract representative of the class of relevant hydrogen atoms (the experimental value is the value found in *the hydrogen atom*; recall again the discussion in 2.1) or not (the experimental value is the value found in all, or all relevant, hydrogen atoms), Weinberg's intention to report an empirical fact is entirely obvious.

We now see how the contact with reality is only alluded to. It is realized only by the comparison implied in the phrase "agreement . . . with the present experimental value." Everything else is left to the able reader. In particular, the reader is assumed to understand that, because the agreement of both values is "excellent," the theory has successfully explained the Lamb shift. So the whole text that develops the explanation is not making reference to entities in the real world—apart from the description of "the experimental value" inviting the final comparison. The step of establishing the contact between the body of the explanation and the explanandum—so trivial a step for the understanding scientist, but so intricate a problem for the analyzing philosopher—is then left to the reader. Especially, also any presumption of QFT being a description of real atoms is left implicit. Hence, a typical scientific theory in a typical situation of its application makes contact with reality only at one point, the endpoint of a long and substantial reasoning. Indeed, the contact is so superficial that on the textual surface no substantial part can be interpreted as referring to reality. (Subtracting the explanandum (i), nothing but the phrase "experimental value" in (iv) refers to any real entity.) A fortiori, the theory involved cannot be read as describing reality.

There is a final complication here. Because of its explanatory success (just exemplified by Weinberg) QFT can indeed be viewed as a description of a crucial physical aspect of the real world, especially of the breaking up of degeneracy in energy levels of light atoms. If, invited by the exposition, the audience takes this attitude toward the theory, it does so after the fact. This attitude may lead to new textbook descriptions where theories are straightforwardly assumed to describe reality. The explanatory (or predictive) power justifying the attitude will only intermittently be commented on. But the attitude may also be taken toward a high-level theoretical text like Weinberg's treatise, after the text has demonstrated the theory's explanatory power. Let's assume that Weinberg's demonstration of the excellent agreement of theory and experiment does itself invite such an attitude toward QFT. The theory then becomes divorced from the concrete text that presents it. Although the latter still contains all the linguistic details we have looked at, and hence expresses a piece of reasoning concerning theoretical entities, the theory presented in the text now has a different status: It is implicitly viewed as a description of reality. It would be here that (as I have briefly hinted at in the introduction) even token expressions of a theory's presentation in retrospect could be judged as changing their referential status. For example, after QFT, because of its explanatory success, has been accepted as describing the energy levels in light atoms in the real world, we might say that even the opening of the explanation in (ii) describes generically the electrons in the shell of light atoms—in the real world. But to assume that this is the case from the start would blur the true referential situation and would leave it mysterious why (ii), like the typical scientific explanation, opens with a theoretical request, not a descriptive sentence.

3. CONCLUSION

I have asked when a scientific theory describes real-world entities. My presupposition was that theories are dynamical entities in the sense that we can take different attitudes toward them while inventing, corroborating or applying them. My claim was that, although successful theories are often taken to describe the real world as one would naively assume, this attitude toward them is not yet present, or if present must be suspended, when a theory's explanatory success is demonstrated. A theory, at this point, does not describe anything real, and thus we have a situation where the theory performs at its best and yet does not describe anything real. My analysis of Weinberg's explanation of the Lamb shift was meant to illustrate all this. The example is really typical for physical theories and science at large. The detachment from reality I tried to point out in a typical theory application presumably will not go away when we look at those portions of a theory where it is developed instead of applied. What I have loosely called a reference to theoretical, instead of real-world, entities is ubiquitous in respectable scientific theories. Especially in theory applications such theoretical entities are introduced and described in order to generate explanations or predictions.

This loose manner of talking about theoretical entities has not been qualified here. Do I really mean to say that theories at crucial stages, rather than describing reality, describe a realm of theoretical entities? I have left open this question. An answer to it would establish a deeper contact with the topic of this volume: the role of fictions in science. Because theoretical entities are not abstracted from observation and do not exist independently from the communication situations where they are introduced, there is no clear distinction between them and fictitious entities. We might as well call them fictitious, subtracting of course the usual association of fiction with uncontrolled arbitrariness. Given this move, we would indeed have seen an illustration of how large portions of a respectable physical theory in action refer to fictitious entities. But accepting this further move ignores the question of whether these references to theoretical entities are cases of real reference, but to fictitious entities, or cases of pretended (fictitious) reference to real entities. The second option is so attractive because it helps us to avoid a messy realm of entities, addressed in scientific theories, that are fictitious (do not really exist) but are really referred to (hence, must in some sense exist). I am much attracted to this option, but making a case for it is not so simple, as the Lamb shift example illustrates. The punchline of Weinberg's text was a demonstration of properties that must be distributed to different entities. To say that one of these entities in no way exists and that we just have pretended to be referring to an entity makes it difficult to accommodate the comparison. For how can we have a substantial comparison if one of its relata does in no sense exist? This problem illustrates most vividly that the question of whether and eventually what a theory describes at stages where it clearly does not describe portions or aspects of the real world deserves further attention.[9]

NOTES

1. The sense of this somewhat cryptic remark will become clear below at fn. 7.
2. This is true if my claim is correct that the atemporal present tense of the verb signals genericity In this case, the generic inflection generally determines the one in the noun phrases to be also generic—as has already been observed by Chafe (1970, pp. 189–190).
3. For this interpretation of generic statements as expressing dispositional properties see again Chafe (1970, p. 169).
4. The idea originates with Suppes's (1962) classic paper; see, e.g., Harris (2003) for an up-to-date exposition and a description of outlier elimination and extrapolation.
5. The classic paper is Prince (1981); see this paper's discussion in Lambrecht (1994, pp. 85–86).
6. What if, contrary to the Weinberg example, the modeling purpose were communicated at once with introducing the theoretical entity? In this case, it becomes difficult to argue that the entity so introduced is not intended to represent anything real, whereas the model is. Let's suppose that in this case the theoretical entity does represent something real, because of the explicit modeling intention. Still, I would argue for a difference between theoretical entity and theoretical model. We do not need to presuppose that a theoretical entity is intended to represent something for us to understand that it is introduced and is attributed this or that property, but we must presuppose so to understand the modeling purpose.
7. It is in (iv) that the explanation might be interpreted as turning into a prediction. The classic Lamb shift is just "of order 1000 MHz" (see (i) earlier) and Weinberg's calculation, concluded in (iii), reproduces the phenomenon after the fact, thus explaining it. Now, is the "present experimental value" (of 1981 and 1986) another version of the Lamb shift to be explained? Well, at least in one sense it isn't. The bulk of the QFT techniques used by Weinberg was developed before the 1980s and so antedates these experimental values. QFT thus predicts these values. (The "agreement" Weinberg refers to is no reliable indicator, as it might be a relation of fit between explanandum and explanans as well as between prediction and testing data.)
8. There is a digression in the middle of the section where Weinberg uses the high-energy part of the energy shift to give a "fair order-of-magnitude estimate of the Lamb shift without further work." To get a numerical result, "the binding energy of the electron in the atom" must be estimated for a light atom (p. 582). Thus, here the Lamb shift, but not the hydrogen atom, is mentioned. An interpretation of this digression would go along the same lines as my interpretation, in the main text, of Weinberg's main result.
9. Thanks to Mauricio Suárez and the participants of the 2006 Madrid conference. Especially, comments by Arthur Fine and Kate Elgin have helped to sharpen the title question of this paper. Thanks also to Christian Lehmann for advice on some linguistic details.

9   Scientific Fictions as Rules of Inference

Mauricio Suárez[1]

1. FICTIONALISM IN THE PHILOSOPHY OF SCIENCE

Hans Vaihinger's philosophy of "as if" introduced fictionalism into philosophical discussions in the early years of the 20th century.[2] Vaihinger's thesis was radical: A knowledge worth pursuing is thoroughly infused by fictional assumptions. Vaihinger distinguished carefully fictions from hypotheses, and considered most of science and mathematics to engage shamelessly in the production, dissemination, and application of both. Hypotheses are directly verifiable by experience and their truth is tentatively granted. Fictions are, for Vaihinger, accounts of the world and its systems that not only are plainly and openly false, but knowingly so, yet remain indispensable in theorizing—in science and elsewhere. However the fictions employed in scientific reasoning are not of the same kind as those that appear in other areas of human endeavor. Vaihinger distinguished scientific fictions from other kinds of fictions (such as poetic, mythical, or religious fictions), and he understood the difference to be one of function. Virtuous fictions play a role in a particular kind of practical rationality in scientific theorizing, a kind of "means–end" rationality at the theoretical level. In Vaihinger's terminology, they are *expedient*.[3] These are the fictions that figure in the scientific enterprise, and among the most prominent throughout the history of science Vaihinger identified forces, electromagnetic "lines" of force, the atom, and the mathematical infinity, as well as some of the main constructs of differential analysis such as infinitesimal, point, line, surface, and space.

Vaihinger's work is unfortunately not sufficiently well known today, but he should appear to philosophers of science as an extremely contemporary figure. A recent brief paper by Arthur Fine brings Vaihinger back to the philosophical fore.[4] Fine notes that "Vaihinger's fictionalism and his "as if" are an effort to make us aware of the central role of model building, simulation, and related constructive techniques, in our various scientific practices and activities" (Fine, 1993, p. 16). There has been in the last decade or two a remarkable resurgence of interest in the topic of modeling, which emphasizes the essential role played by idealizations, contradictions, abstractions, and simulations in the practice of scientific modeling.[5] Vaihinger's work fits right in: His concern to emphasize the use of false or contradictory assumptions in building models of systems and their workings is of a piece with this whole body of literature. So too is his concern to appreciate the pragmatic virtues that these assumptions might bring to scientific reasoning. For instance, Vaihinger also distinguished between *fictions* (which involve internal contradictions, or inconsistencies) and what he called *semi-fictions* (which involve contradictions with experience).[6] We would nowadays refer to the former as contradictions, and claim that they are logically false, while calling the latter idealizations, which we would claim are empirically false. This distinction will be appealed to freely in the text, and instances of both types of fiction will be identified. But I will not be concerned very much with the fact that fictions in science tend to entail falsehood. Although Vaihinger emphasized the falsehood of scientific fictions, his main concern was with their cognitive function.[7] The main interest throughout this chapter is in shedding some light on the cognitive function of scientific fictions, regardless of their truth value. In particular I urge that expediency in inference is the main defining function of a scientific fiction. This chapter is a first attempt at an elaboration and defense of this inferential expediency of scientific fictions. It can be seen as part of a larger argument for taking the cognitive value of a scientific fiction to lie entirely in its function in inquiry, and to be fully independent of its truth value.

The modeling scholarship of the last decade or so has made a strong case for a version of Vaihinger's main thesis: The use of fictions is as ubiquitous in scientific narratives and practice as in any other human endeavor, including literature and art; and scientists have demonstrated throughout history a capacity to create, develop, and use fictions for their own purposes that compares with that of any writers or artists. This is why this assumption will not be defended here, but will rather be taken as a matter of fact, and a starting point. The aim is instead to explore a philosophical issue that is key to the success of the philosophy of modeling movement, and to Vaihinger's conception in particular. This is the fundamental distinction between what Vaihinger identified as *scientific* fictions and other kinds of fictions.[8] It is this distinction that lies at the heart of the thesis that fictions are of particular use to science, or that science employs them in a particularly useful way. Without an elaboration and defense of the expediency that characterizes scientific fictions the main thesis, however true, would bear little content—and it would shed little insight on scientific practice.

2. FICTIONS IN THE HISTORY OF SCIENCE: TWO EXAMPLES

In order to characterize scientific fictions we may begin with some illustrious examples of productive use of fictions in the history of science, the ether theories and the models of the atom at the turn of the century. They are interesting in complementary ways. In the ether case the use in mathematical and experimental inference of what we nowadays take to be a fictitious entity was extremely productive and long-lasting—although convictions as to its reality differed and wavered considerably. The case of atomic models illustrates another possibility: a model of a putatively real entity that was never taken very seriously—the model was never assumed to be other than a fictional description of the entity—and was consequently short-lived, yet turned out to be extraordinarily useful as a heuristics for developing new, more detailed and powerful models. Therefore these cases illustrate the difference between representations of fictional entities and fictive representations of real entities; they also illustrate the difference between long-lasting formal tools and short-lived concrete models. Yet Vaihinger's thesis holds in both cases, because fictions are involved in an essential way, one way or another. The description that follows will emphasize some features that in both cases seem relevant to the distinction between scientific and nonscientific fictions. (Ultimately the philosophical discussion will focus on a further example from quantum theory.)

Throughout the 19th century the ether was taken to be the putative carrier of the light waves, and stellar aberration phenomena had established that the earth must have been in motion with respect to it. The most sophisticated mathematical theories of the ether were developed in Britain in the wake of Maxwell's theory of electromagnetism. Maxwell developed his theory in the years 1856 to 1873, the year of publication of his *Treatise on Electricity and Magnetism*.[9] The theory went roughly through three phases corresponding to the publication of "On Faraday's Lines of Force" (1856), "On Physical Lines of Force" (1861), and finally "A Dynamical Theory of the Electromagnetic Field" (1865) and the *Treatise*. In each step Maxwell's theoretical model acquires a higher degree of abstraction from the mechanical model of a luminiferous ether. The central concept throughout is that of field energy: the only concept toward which Maxwell shows an increasingly firmer ontological commitment over the years. By contrast, Maxwell's reluctance to accept the ontological implication of the existence of the mechanical ether is strong and explicit from the beginning. As Morrison has pointed out, in drawing the analogies in "On Faraday's Lines of Force" between electrostatics, current electricity and magnetism with the motion of an incompressible fluid, this fluid "was not even considered [by Maxwell] a hypothetical entity—it was purely fictional" (Morrison, 2001, p. 65). And later on as he developed a clearer view of what have come to be known as Maxwell's equations, Maxwell remained resolutely skeptical regarding the existence of the ether. The mechanical models of the ether were gradually stripped of ontological content as the dynamical equations were developed, yet these mechanical models remained indispensable to scientific theorizing, according to Maxwell, "as heuristic devices, or at best, descriptions of what nature might be like" (Morrison, 2001). In other words, the ether remained for Maxwell a fiction—with an essentially heuristic function in the development of increasingly sophisticated mathematical models that captured the empirical phenomena. This was its scientific virtue—to be a guide in the construction of more detailed, explanatory or predictive models of the phenomena.[10]

The Maxwellian tradition was continued among theoreticians in Cambridge, mainly through the work of Charles Niven and Joseph Larmor. Andrew Warwick has written a scholarly and detailed history of the Cambridge couching in electromagnetic theory from the very first lectures that Maxwell himself gave there on publication of his *Treatise* until the final demise of ether-based theories in favor of Einstein's theory in the 1920s (Warwick, 2003). What is remarkable about this history is the role that the ether played as basis for the application of electromagnetic theory to all kinds of practical and experimental problems. Although the Cambridge Maxwellians professed a belief in the ether (and thus, unlike Maxwell himself, took the ether to be a hypothesis rather than a fiction in Vaihinger's terms), they nonetheless employed the ether as a mental construction that allowed them to apply electromagnetic theory in a much simpler and straightforward way. The critical difference is that the ether theories allow for electromagnetic effects to be present all over as due almost entirely to the flow of energy "in the ether," while action-at-a-distance theories supposed that electromagnetic effects only occurred within conducting and dielectric material, not at all in empty space. As Warwick puts it, "the solution of problems using the field-theoretic approach therefore required very careful consideration of the electromagnetic action and boundary conditions applicable at or near the surface of conductors" (Warwick, 2003, p. 329). Confidence in the applicability of the equations is required for carrying out these calculations, but a belief in the existence of the ether is not strictly required. In fact, convictions as to the existence of the ether wavered a great deal, and differed between its different proponents.[11]

In other words, the fiction—or hypothesis—of the ether has for both Maxwell and Larmor a powerful positive heuristic for further research, giving guidance on how to generate more detailed models of the phenomena, allowing quick and effective inference of experimental and practical results that can then be readily checked against experience. It provides an excellent tool for the calculation of effects and the regimentation of knowledge. It also provides what Mary Hesse aptly called neutral analogies—unproved similarities between the ether models and real phenomena that call for exploration and investigation.[12] But the reality of the ether itself is not required for inference, taxonomy, or analogy. The ether provides a mental model of great expediency and ease of calculation, but it can well remain "a mere figment of the mind, not a fact of nature" (Maxwell, on referring to disturbances of the ether, as quoted in Morrison, 2000, p. 96).

The second example is even more startling in the fictitious nature of the description employed: J. J. Thomson's *plum pudding* model of the atom. Although the entity described is by no means considered a fiction by our present lights, the description of the entity given in the model in question certainly is fictional and in ways interesting to our present purposes.[13] The idea that electric conductivity phenomena may be underpinned by the fundamentally asymmetric nature of the distribution of positive and negative charges in atoms actually goes back to Lord Kelvin, who proposed it in his President's Address to the Royal Society in 1893. In fact, Kelvin had already proposed something very similar to a plum-pudding model for the atom as part of his defense that Crookes's experiments on cathode rays established that they were negatively charged particles. Thus Thomson did not "make up" the plum pudding model, but was following in Kelvin's footsteps in proposing and developing it in the years between 1897 and 1910. The model was part and parcel of the strategy followed by both Kelvin and Thomson in trying to prove the particulate nature of cathode rays, and the fundamental electrical asymmetry of conductivity phenomena.

According to the plum pudding model the atom is a roughly spherical *sponge* formed by evenly distributed positive charge in which minute negatively charged particles ("electrons") are inserted, like raisins in a traditional British Christmas cake. The model had the great advantage of explaining ionization phenomena, whereby negative charged particles are bounced off atoms by collisions with other atoms, creating electric currents. It also explained the production of cathode rays and sustained Kelvin's and Thomson's favored interpretation of them as negatively charged particles. But most importantly, the plum pudding model was a powerful heuristics in the development and application of Thomson's "working hypothesis" between 1897 and 1910 for the experimental inquiry into electrical phenomena in gases, namely, "that the negatively charged corpuscle is universal and fundamental, ionization results from the dissociation of a corpuscle from an atom, and electrical currents in gases at low pressures consist primarily of the migration of corpuscles" (Smith, 2004, p. 23).

As is well known, the plum-pudding model was refuted by the experiments performed in 1909 by one of Ernst Rutherford's collaborators in Manchester, Hans Geiger, together with a student, Ernst Marsden.[14] In these experiments thin foils of gold were bombarded with alpha particles (essentially a couple of protons and neutrons bound together as in helium nuclei) and a significant proportion of recoil was observed. (On the order of 1 in 20,000 alpha particles was deflected by an average angle of 90 degrees.) The effect is sometimes known as Rutherford scattering, and was effectively employed by Rutherford[15] in order to reject the plum pudding model because the electrons in the atom are too small and light to produce the recoil effect, while the diffuse "sponge" would be porous to the massive and energetic alpha particles. Hence Rutherford advanced his hypothesis of the existence of a massive and discrete nucleus at the center of empty space, orbited by minute electrons, in order to explain the scattering effect with the experimentally observed probability.

It would be uncharitable, however, to charge Thomson with the belief that his plum pudding model was a complete and true description of the atom (see Smith, 2004, particularly p. 25). Instead the model served three different heuristic functions: (i) it allowed Kelvin and Thomson to argue for the particulate as opposed to wave-like nature of cathode rays (electrons), a fact that Thomson's 1897–1899 papers were instrumental in establishing universally; (ii) it justified and provided motivation for Thomson's "working hypothesis" regarding conductivity in gases during the 1897–1910 years, which in turn led to many important experimental results and conclusions—including among others the establishment of the existence of electrons themselves; and (iii) it helped Rutherford target the experimental evidence available toward the missing or defective assumptions in the previous models, thus enabling him to discover the existence of the nucleus. (As is well known, Rutherford's "planetary" model was in turn superseded by Bohr's early quantum model [1913], which hypothesized spontaneous quantum transitions between electron orbitals, and hence a nonclassical structure of quantum energy levels within the atom.)

It is hard to overestimate the importance of the plum pudding model in helping focus inference making in all these three different areas. The historical facts rather point out that these three crucial and critical developments and discoveries in the history of modern physics could hardly have come about without the expediency in reasoning provided by the plum pudding model, no matter how fictitious. The model served an essential pragmatic purpose in generating quick and expedient inference at the theoretical level, and then in turn from the theoretical to the experimental level. It articulated a space of reasons, a background of assumptions against which the participants in the debates could sustain their arguments for and against these three hypotheses. As Thomson himself put it: "My object has been to show how the various phenomena exhibited when electricity passes through gases can be coordinated by this conception."[16]

3. FICTIONS IN CONTEMPORARY PHYSICS

The use of fictions is not confined to past or failed science. Let us now consider a couple of typical instances of fictions within contemporary successful science. First, an example from astrophysics is briefly discussed: the Vaihingerian semi-fictions involved in models of stellar structure. Then we turn to a striking case of Vaihingerian full fiction: the models of the theory of quantum measurement. Once again these case studies emphasize the inferential function of the fictional assumptions involved in both models.

3.1. The Semi-Fiction of a Star

Models of stellar structure in astrophysics provide a description of the inner workings of a star; in particular the fuel burning processes (nuclear fusion) that turn hydrogen into helium and generate the star radiation, while accounting for the star life cycle and evolution. These models match up the observational quantities of a star, which mainly pertain to the properties of its photosphere, that is, the outermost layer of the star. The observable quantities of stars include: (i) its luminosity (energy radiated per unit time, which depends upon apparent brightness and distance), (ii) its surface temperature (the temperature of the photosphere), (iii) the photosphere's chemical composition, and in rare cases (binary stars) the mass of the star. The models allow us to infer conclusions regarding the internal workings of a star on the basis of these observational quantities. Their cognitive value depends upon the reach, power, and ease of calculation of such inferences.

These models make at least four assumptions that are widely assumed to contradict either the physics of matter and radiation, the physical conditions of the interstellar medium, or both. Hence the models are knowingly strictly speaking false of real stars. I won't discuss them here in detail, but will just briefly discuss these four assumptions (see, e.g., Prialnik, 2000, pp. 6–8. Also Tayler, 1970). First, models of stellar structure assume that a star is an isolated bubble of gas; that is, they assume that a star is a physically closed system—no external forces (gravitational or electromagnetic) intervene. The only forces that can affect the internal structure of a star are consequently assumed to be the star's internal gravitational forces due to the rotational movement of the gas, and the forces that arise out of the nuclear burning inside the star. As far as external gravitational forces go, the assumption of isolation is not entirely unrealistic: A star is a dense concentration of hydrogen in the interstellar medium, and the nearest such concentration can be on average as far as 4.3 light years away (this is roughly the distance between the sun and Alpha Centauri, the star nearest to our sun). So external gravitational forces are bound to be tiny, and therefore negligible for all practical purposes. But the assumption of isolation is physically quite incorrect: The insterstellar medium is not empty space but is replete with irregularly dense (although on average much lighter) gas, mainly hydrogen, which in the surroundings of the star is gravitationally attracted into the star, while radiatively repelled. In other words, isolation suggests that the physical boundaries of a star are sharp, when as a matter of fact they are rather imprecise.

Second, stellar structure models assume that all stars possess identical chemical composition, namely, the sun's: 70% hydrogen and 30% helium. The presence of heavier elements has been confirmed in every star, but it is essentially irrelevant to the structure of the star in most hydrogen-burning models, so it is entirely neglected. In these models, a small difference in the proportion of hydrogen and helium in the initial composition of a star can have an important effect in its evolutionary life (it can particularly affect its luminosity and lifetime), but because the models have other parameters to adjust for these observable quantities, these differences are essentially ignored, and a blanket assumption is made.

Third, it is assumed that the shape of the star is spherically symmetrical throughout its life. Yet both internal rotational forces and magnetic forces are well-known causes of departures from spherical symmetry. Hence this assumption rules out such internal forces, which as a matter of fact are known to be large. Finally, a star is assumed to permanently stay in a state of thermal equilibrium, whereby the temperature of the gas is identical to the temperature of the radiation emitted (see Prialnik, 2000, p. 16, Chapters 3 and 5). The implication of this assumption is that radiation provides the only form of energy transfer within a star. It also follows that the energy spectrum of a star is a black-body spectrum. Yet, massive convective fluxes are known to occur in periods of heavy hydrogen burning within the star. So the assumption is that such convective forces have no impact on the temperature distribution inside the star. Because the inside of a star is entirely theoretical, the only justification for this assumption is the model's ability to correctly match the observable quantities of the photosphere.

These assumptions are not discussed any further here. I merely emphasize that they are all known to be empirically false, or at any rate unproven, yet in combination they afford a huge improvement in the expediency of the inferences that can be drawn from the models to the observable quantities, and between such quantities Their effect on the accuracy of predictions from the model cancels out, and is therefore negligible in comparison. In other words, the stars of contemporary astrophysics' models are Vaihingerian semi-fictions whose justification lies entirely in the great ease and expediency in inference making that they generate.[17]

3.2. The Full Fiction of Quantum Measurement

An example from present-day theoretical science is next discussed that exhibits the characteristic expediency of fictions in a prominent and illustrative manner. The model that quantum theory provides for measurement interactions is intended as a representation of a physical process—an image of a physical process provided by a highly formal and mathematical model. It might seem surprising that I am characterizing it as a fiction, because the formal machinery of quantum theory is so solidly entrenched among practicing physicists. And yet, its fictional character has actually been proved as a mathematical theorem—known as the insolubility proof of the measurement problem. First the basic assumptions underlying the model are discussed (sec. 3.2.1), and then reasons are provided for considering it a semi-fiction in Vaihinger's sense (sec. 3.2.2), a maximal semi-fiction, and a fully fledged fiction (sec. 3.2.3). Sections 3.2.1 and 3.2.2 are technically a little demanding (even though the most demanding technicalities have been confined to an appendix), and the uninformed reader may well want to skip them without loss.

3.2.1. The Quantum Theoretical Model of Measurement

Quantum theory provides an abstract mathematical model of physical inter-action between a microscopic quantum object and a macroscopic device designed to test the state of the object. The theory, as first formulated by von Neumann (1932), ascribes a quantum state to the measuring device, and treats the interaction as a quantum interaction, that is, one that obeys the Schrödinger equation.

Suppose the initial state of the system is $W_o = \Sigma_n p_n P[v_n]$, where each v_n may be expressed as a linear combination of eigenstates of the observable O of the system that we are interested in (i.e., $v_n = \Sigma c_i \phi_i$); and that the initial state of the measuring device is $W_a = \Sigma_n w_n P[\gamma_m]$. Throughout the chapter I refer to the observable represented by the operator $I \otimes A$, as well as that represented by A, as the pointer position observable. The eigenvalues of this observable are the set $\{\mu_n\}$. As the interaction between the object system and the measuring device is governed by the Schrödinger equation, there must exist a unitary operator U that takes the initial state of the composite system (object system + measuring device) into its final state at the completion of the interaction, as follows: $W_o \otimes W_a \rightarrow U (W_o \otimes W_a) U^{-1}$. (For further details of the interaction formalism, see Appendix 1.)

3.2.2. The Problem of Measurement

The intuition behind the so-called "problem of measurement" is easy enough to state. Take a system in an arbitrary superposition $v_n = \Sigma c_i \phi_i$. Then, due to the linearity of the Schrödinger equation, at the conclusion of an ideal measurement interaction with a measurement apparatus in any pure state, the composite (system + device) will be in a superposition of eigenstates of the pointer position observable. And according to the so-called eigenstate–eigenvalue link (e/e link), the pointer position observable cannot have a value in this state, because it is not in an eigenstate of the relevant observable. But surely quantum measurements do have some values—that is, they have some value or other. Hence the quantum theory of measurement fails to describe real quantum measurements, and the model expresses a fiction—to be exact, a *semi-fiction* in the terminology of Vaihinger, because it contra-dicts empirical reality.

3.2.3. The Model is a Maximal Semi-Fiction

Let us refer to a model as a maximal semi-fiction in Vahinger's sense if every assumption in the model can be shown to be false. Then the quantum theoretical model of measurements is a Vahingerian maximal semi-fiction.

All the assumptions required for a quantum theoretical model of measurements are strictly speaking false. In addition to the eigenstate–eigenvalue link and the assumption that the Schrödinger equation is the full dynamical description of events, there are two formal conditions that are required to derive the problem of measurement. I have elsewhere referred to them as the *transfer of probability condition* or (TPC), and the *occurrence of outcomes condition*, or (OOC). (They are both described formally in Appendix 2.) Informally, (TPC) states that the probability distribution over the eigenvalues of the initial object system should be reproduced as the probability distribution of the pointer position observable. (OOC) by contrast states that the final state of the composite is a mixture over states in which the pointer position observable takes a particular value or other with probability one, and is often thought to be inspired by the eigenstate–eigenvalue link.

(TPC) is strictly speaking false on at least two counts. First, it assumes that whether interactions are measurements is an all-or-nothing affair that does not depend on the actual initial state of the system to be measured at a particular time, but on all the possible states that the object may have had in accordance with the theory. This is hardly satisfied by any real measurement we know. For instance, in setting up a localization measurement of the position of an electron in the laboratory, we do not assume that the device should be able to discern a position outside the laboratory walls, even if it is theoretically possible that the particle's position be infinitely far away from us. All real measurement devices are built in accordance to similar assumptions about the *physically* possible, as opposed to merely theoretically possible, states of the object system, on account of the particular conditions at hand. So real measurement devices do not strictly speaking ever fulfill (TPC).

Second, (TPC) appears to require measurements to be ideal in the technical sense of correlating one-to-one the initial states of the object system with states of the composite at the end of the interaction. However, many real measurements are not ideal in this sense. Most measurement apparatuses make mistakes, and no matter how much we may try to fine-tune our interaction Hamiltonian, we are likely in reality to depart from perfect correlation. So again, in most cases of real measurements (TPC) will not apply.

Let us now turn to (OOC). This is also strictly speaking false, because it assumes that the measuring device can only "point" to the eigenvalue of the pointer position observable that has probability one in the final state that results at the end of the interaction. But we can see that this assumption is false in most measurement interactions with quantum objects in mixed states where outcomes are produced, and pointers "point" in spite of the probabilistic nature of the transitions.

Finally, the application of the Schrödinger equation in this setup is also strictly speaking false because it implies the assumption that all quantum systems, not only composite systems involving measuring devices, are closed systems. It assumes that the quantum Hamiltonian can transform pure states into pure states, or mixtures into mixtures, but never a pure state into a mixture or vice versa. In reality, all systems are open and subject to a degree of state shift due to interaction with the environment and background noise—this phenomenon is known as decoherence.

3.2.4. The Model Is a Full Fiction

But in fact the insolubility proof of the quantum measurement problem shows that these three premises (together with a fourth premise named RUE, which I have ignored here) are inconsistent under the standard interpretation of quantum observables, the so-called eigenstate-eigenvalue link, or e/e link.[18] Hence the model is not only empirically false, and maximally so, but it turns out to be *necessarily* false, because it is internally incoherent. In Vaihinger's terminology, the quantum theoretical model of measurements is a *fully fledged scientific fiction*.

3.2.5. The Model Is a Scientific Fiction

The main feature of the model, shared with most theoretical models, is its inferential capacity. Once we understand how the model works we are in a position to draw inferences regarding the typical final state of a composite (object + measuring device) after the interaction. From this state we can further infer the possible values of the pointer position observable of the measuring device at the conclusion of the interaction (as long as the final state has a particular form that allows us to do just that). And from this inference we finally come to understand that there is an inconsistency in the account, which in turn leads us to consider which among the assumptions might be false.

Similarly, we have a large array of applications of measurement theory to different cases of physical interactions at the quantum level, each of them relaxing different assumptions of the model. For example, by relaxing the (e/e link) we have models within the modal interpretation of quantum mechanics; by relaxing the assumption of Schrödinger evolution of the composite we derive stochastic reduction and quantum state diffusion models; by relaxing the (OOC) assumption we obtain statistical interpretations of the measurement interaction, and hidden variable models. Finally, relaxing (TPC) allows us to encompass and develop models of highly nonideal interactions, such as destructive measurements.[19]

As in the previous example, the fictitious entity or process—the *ether* described by ether theories, the atom as plum-pudding, the internal structure of stars, measurement interactions as described by the quantum model—has in the appropriate formal model the capacity to generate various inferences to experimental or practical results, some of which can then be tested against experience. It is not the fiction itself that is in the first instance under experimental test, but the results of inferences that the fiction licences in appropriate formal frameworks, or in conjunction with further assumptions and background theory. In all these cases the inferences would either be arbitrary or impossible to even draw without the supposition of the fiction in the first instance. The fictitious entities or processes articulate a framework for quick and expedient inference-making that would be either impossible or arbitrary otherwise.

4. A SUBSTITUTIVE FUNCTIONAL ACCOUNT OF SCIENTIFIC FICTIONS

On the account defended here the hallmark of scientific fiction is *expediency in inference*. Note that this is not equivalent to the claim that the fictions employed in art and literature, unlike those from science, do not posses a capacity to generate useful (imaginative, pleasant, interesting) inferences. It is obvious that literary and artistic fiction can and must serve that purpose too. But expediency is not generally considered a virtue. To the contrary, it is often derided for good literary fiction. Nor is it ever required in such cases that the conclusion of our inferences be at least in principle testable against experience. There is thus a double norm that rules the use of fictions in science in comparison with nonscientific fictions. First, scientific fictions will be judged by their capacity to allow expedient inference-making. Second, at least some of the conclusions arrived at by means of such inferences will be taken to be empirically testable at least in principle. Most fictional assumptions in science will possess both functional virtues. By contrast, the fictions of art and literature can at best share in the first kind of virtue, but they need not even do so in order to fulfill their proper aesthetic function.

So how do these "fictitious" representations in science work? There is by now a large literature on the topic of representation, particularly as applied to artistic or pictorial representation. A comparison with such theories will be illustrative. The use of fictions in science seems to share some elements in common with Gombrich's substitutive account of fictions in art, and his account of pictorial representation more generally.[20] On this account, the function of pictures is to cognitively replace the objects that they depict—the replacement has a cognitive function in the sense that it allows surrogate reasoning. In reasoning about the properties of the picture, we may be able to infer properties of the object depicted. Gombrich employs the analogy of a hobbyhorse, which children play with in a similar substitutive fashion. In this example, the replacement is not merely cognitive, but also of a practical nature (often missing in the painting and in the scientific case), so actions can be performed around the hobbyhorse that would be performed around a real horse. In these activities children can sometimes lose track of the fictional nature of the entity—in fact there is a sense in which for it to perform its function correctly, it is essential that the fictional nature of the entity be in some ways suppressed. Although in Gombrich's substitutive account of representation we are not required to gain the (false) belief that the fictitious entity itself exists,[21] it seems that we are required to display at least some attitudes toward the fiction that we would display toward the real entity. In the hobbyhorse case this "pretence" attitudinal set includes practical action; in the painting case it merely includes cognitive factors.

Something very similar operates in the case of scientific fictional representation. What we do—what scientists do—when employing a fiction for scientific purposes—what Maxwell does in employing the ether, Thomson in employing the plum pudding model, or quantum physicists in applying quantum measurement theory—is to substitute the real process (with all its complications, disturbing factors, and exogenous causes) with a simpler, streamlined, and coarser-grained fictional account (one where all the factors and causes are fully and carefully identified), which is more expedient in the practice of reasoning and inference-making. We then go on to investigate the properties of the substitute rather than the real process.[22] It will be typical for many scientists engaged in that process to lose track of the fact that the entities invoked are fictitious. This is as it should be on a substitutive account, because the representational success of the model depends on its success in generating the same set of cognitive attitudes toward it that scientists would exhibit toward the entity modeled.[23] The attitudes of Larmor and Trouton toward Maxwellian ether-based electro-magnetism just reflect this: Although there was never any experimental confirmation of the ether (and Larmor himself thought such confirmation was made impossible by the theory itself—i.e., by the ether-fiction itself), a realist attitude toward the ether assumption was a powerful heuristic in applying the theory to the most diverse phenomena. But although a realist cognitive attitude will be typical, a belief in the fiction is not necessary in order to operate the model successfully—we already saw that Maxwell himself displayed a much more cautious attitude toward the ontological import of the ether.

And *mutatis mutandis* in the quantum measurement case: To model quantum measurements in this way involves the cognitive attitude toward the model that we would adopt if confronted with real measurement processes. Thus it becomes possible to draw inferences regarding the real outcomes of real measurements from the model. But a belief that real quantum measurement interactions fully obey the formal assumptions is not required for such a cognitive attitude, which can be displayed simultaneously with a belief in the fictitious character of the entities or processes involved. (In fact, such a belief is, in the measurement case, impossible on pain of inconsistency and irrationality—as is shown by the insolubility proofs.) Thus the cognitive attitude that is needed to apply and use scientific fictions is pragmatically indistinguishable from belief, but it is *not* belief.

5. REPRESENTATION AND FICTION-BASED INFERENCE: AGAINST SIMILARITY AND ISOMORPHISM ACCOUNTS

Many of the most highly theoretical and mathematical models employed by scientists either involve fictional assumptions in their representations of real systems, or otherwise they refer, or purport to refer, to fictitious entities or processes. The representational character of these models can be explained roughly along the lines of a substitutive model of pictorial representation, which explains why the main cognitive function of these models is surrogate reasoning regarding its objects. This requires a "realist" attitude toward the model only in the sense of surrogate reasoning and inference, and as a powerful heuristic for further research. It does not require the belief that the model is an accurate representation nor that the entities described in the model are in fact real, and in fact it often will be inconsistent with such a belief.

In this final section I argue that this representational function of fictitious models in science is best understood by reference to a broadly construed "inferential" conception of scientific representation, such as the one proposed in Suárez (2004b). On this conception we can say of a model A that it represents an entity, system, or process B in case (i) the representational force of A is B, and (ii) A allows competent and informed agents to draw specific inferences regarding B. A virtue of this account is that it presupposes no relation of reference or denotation between A and B. The notion of representational force is defined so that it is fulfilled by any attempt at reference or denotation, however unsuccessful, that accords to the social practices and norms conventionally adopted in the use of such representational force. Also the notion of "inferential capacity" is fulfilled by any model that has sufficient internal structure to permit inferences regarding its "target," regardless of whether it denotes it, or indeed regardless of whether it is intended to denote it.[24]

One of the advantages that I have claimed for this account is its ability to provide the most natural account of fictional representation.[25] On this account there is no fundamental distinction in kind between a representation of a real entity or process, and a representation of a fictional entity (other than the obvious one of the existence of the entity so represented). Both are representations in virtue of fulfilling exactly the same conditions, and the nature of the representation is the same in both cases. Recall that according to this model there can be representation without reference: The existence of the target B is not required for conditions (i) and (ii) to obtain. In such cases we shall want to ask what the scientific interest of the inferences carried out under (ii) might be—and in most cases there will be inferences to true conclusions regarding observable aspects of the phenomena. Hence ether models allow a good deal of accurate predictions regarding the optical properties of light, and the plum pudding model allows, as we already saw, plenty of new and interesting predictions regarding atomic observed phenomena; stellar structure models allow us to predict the correlations between the observable surface properties of the star, and so on.

In both the ether-based theories and the atom models cases, conditions (i) and (ii) are fulfilled. Maxwell's theory was widely accepted to represent the electromagnetic configuration of an all-pervading medium capable of transmitting energy. As we saw, this was accepted by Maxwell himself, in spite of his rather skeptical views on the existence of the ether itself. The remarkable subsequent Cambridge scholarship in applying electromagnetic theory described in Warwick's book demonstrates that condition (ii) is also satisfied. Huge amounts of effort, time, and energy are spent in the task to determine the inferential consequences of the ether models. Similarly, the plum pudding model is a model of atomic structure—and this is its representational target. Its empirical refutation by Rutherford shows conclusively that condition (ii) is also satisfied—because it shows that it is possible to test experimentally some of its consequences.

Stellar structure models satisfy (i) and (ii) trivially, because they represent real stars and allow very efficiently for inference regarding some of their properties. The quantum theoretical model of measurement interactions satisfies both conditions too—in spite of the internal inconsistency revealed by the insolubility proof. First, nobody doubts that it is a model, or representation, of measurement interactions at the quantum level—and second, condition (ii) is also fulfilled by the diverse applications (which I mentioned in section 2) of different combinations of assumptions from the quantum formalism in the different interpretations of quantum mechanics.

In addition, the inferential conception explains the substitutional character of fictive representations in science that I described in the previous section. The inferential conception denies that substitution is the essential constituent relation of representation, because it claims that no relation between A and B is *required* for representation. Hence it denies that "A is a substitute for B" is necessary and sufficient for A to represent B. However, the inferential account accepts that substitution *à la* Gombrich can be a means of the representational relation.[26] That is, it accepts that this might be the mechanism or relation employed by scientists in surrogate reasoning about B on the basis of A.[27] It thus explains why substitution plays a representational role in these cases.

The inferential conception was proposed in response to the deficiencies of other proposals for understanding scientific representation, namely, the isomorphism and similarity accounts.[28] I believe we have here yet one more reason to adopt it in the face of its competitors, because neither isomorphism nor similarity can provide a similar explanation of the substitutional character of fictive representation in science. Take first the case of what we may call *fictional representation*, that is, representation of a nonexistent entity, such as the case of the ether. According to the isomorphism account, A can only represent B if they share their structure. But in the case of fictional representation this is either false or an empty truism, for it seems impossible to ascribe structure to a nonexistent entity. If the ether does not exist it can not possess any real structure, so isomorphism can not obtain. There might arguably be also a sense in which the ether (a fictional entity) can be ascribed "fictional" structure, but in this sense it can not fail to have the structure that the theory ascribes to it, because it is fully defined by the theory. This points out a significant difference between fictional representations in science and elsewhere: The (fictional) properties of fictional entities represented by scientific models are implicitly defined by the models themselves, which need not always be the case with fictional representation in art or everyday life (for instance, some pictures of Santa Claus may be said to *misrepresent* him).

The same argument seems to apply straightforwardly to the similarity case. Suppose that everything there is about representation in science can be understood through the similarity account; that is, suppose that it were true that A represents B if and only if A and B are similar. Then if similarity requires the sharing of actual properties between A and B, there can be no "fictional" representation in science because B lacks any real properties. We cannot represent the ether by means of models that share actual properties with the ether if there is no ether. If on the other hand it is permissible to ascribe fictional properties to fictional objects and speak of a similarity between a real and a fictional object in virtue of a (fictional) sharing of actual properties—or an actual sharing of fictional ones—then similarity is automatic in cases of the fictions employed by science. It is built into the definition of the entity represented by the theoretical model that it (fictionally) has the properties ascribed to it in this model.

Now let us take the rather different case of the plum pudding model of the atom. What we have here is arguably a very different case of misrepresentation by a deficient model of a real entity *simpliciter*. Let us refer to this type of misrepresentation as *fictive* representation. The similarity conception fares a bit better here because it can be used to characterize the degree and respects in which the model differs from the entity modeled. (Version of the isomorphism conception can arguably do this too.) However, these conceptions can crucially not *explain* the heuristic power of the model, which seems by and large unrelated to any actual similarity or isomorphism between representational sources and targets. Of the three respects in which the model acts a heuristic guide to research (see section 2) these conceptions can only account for the third: Rutherford's experimental refutation of the model. The other two respects in which the model proved heuristically powerful turned out to have nothing to do with any similarities between plum puddings and atoms, nor any isomorphisms between the mathematical structure of plum puddings and that of atoms.[29]

Mutatis mutandis, on either account, for the case of quantum measurement interactions. Whether there are genuine physical interactions between microscopic entities and measurement devices might turn out to be a question of interpretation of the theory (in some versions of the many worlds or many minds theories the notion of genuine physical interaction arguably fails to play a role). In any case, similarity and isomorphism will not shed much light on the heuristic power of a fictional representation of a nonexistent process, nor on the fictive representation (i.e., misrepresentation) of a real one. These practices of fictive and fictional representation may at best be accommodated within these conceptions, but the heuristic value associated with such practices cannot be explained.

Yet in the inferential account, representation is neither false nor trivial in any of these cases. This is straightforward in cases of fictive representation. But then take the harder cases of fictional representation, such as Maxwell's mechanical ether. In the inferential account, representational force could have failed to obtain (and indeed in the present-day, post-Einstein understanding of classical electromagnetism, this theory is no longer intended to represent features of the ether), and the models could have been mathematically so complex as to be completely useless for inference-making and surrogate reasoning—this would be the case for instance if the equations were not analytically solvable. Hence on the inferential conception the substitutive use of the virtuous fictional entities employed in science is a genuine achievement, which requires huge creativity, deep knowledge of the science, and a considerable cognitive ability in the use and development of sophisticated mathematical models.

6. CONCLUSIONS

The substitutive character of the fictions employed in science is best understood by means of an inferential conception of scientific representation. Although cases of fictional and fictive representation may be accommodated within the similarity or isomorphism accounts, the heuristic value of such kinds of representation cannot be convincingly explained. In other words, the isomorphism and similarity conceptions do not elucidate the distinction so dear to Vaihinger and to his present-day followers between scientific fictions and other kinds of fictions. According to Vaihinger this distinction can only be drawn at the level of cognitive function, and I have argued that the inferential conception of representation draws it precisely at the right level.

APPENDIX 1: THE INTERACTION FORMALISM

In this appendix I describe the tensor-product space formalism provided by the quantum theory of measurement to represent the interaction between an object system and a measuring device. Given two Hilbert spaces, H_1 and H_2, we can always form the tensor-product Hilbert space $H_{1+2} = H_1 \otimes H_2$, with dim $(H_1 \otimes H_2) =$ dim $(H_1) \times$ dim (H_2). If $\{v_i\}$ is a basis for H_1 and $\{w_j\}$ is a basis for H_2, then $\{v_i \otimes w_j\}$ is a basis for H_{1+2}. Similarly, if A is an observable defined on H_1 with eigenvectors $\{v_i\}$ and eigenvalues a_i, and B an observable on H_2 with eigenvectors $\{w_i\}$ and eigenvalues b_j then A $\otimes$ B is an observable on H_{1+2} with eigenvectors $v_i \otimes w_j$, and corresponding eigenvalues $a_i b_j$.

Consider two systems S_1 and S_2. If S_1's state is W_1 on H_1, and S_2's state is W_2 on H_2, we can represent the state of the combined system S_{1+2} as the statistical operator $W_{1+2} = W_1 \otimes W_2$ acting on the tensor-product Hilbert space H_{1+2}. If either W_1, W_2 is a mixture, then W_{1+2} is also a mixture. If, on the other hand, *both* W_1, W_2 are pure states then W_{1+2} is pure. Suppose that $W_1 = P_{[\psi]}$, and $W_2 = P_{[\phi]}$, where $\psi = \Sigma_i c_i v_i$ and $\phi = \Sigma_j d_j w_j$. Then $W_{1+2} = \Sigma_{i,j} c_i d_j v_i \otimes w_j$, which is a superposition of eigenstates of A $\otimes$ B in H_{1+2}. More specifically, if S_1, S_2 are in eigenstates of A,B, the combined system S_{1+2} is in an eigenstate of A $\otimes$ B. If $W_1 = v_i$ and $W_2 = w_j$, then $W_{1+2} = v_i \otimes w_j$, a so-called *product state*.

For an arbitrary (pure or mixed) state W_{1+2} of the combined system, and arbitrary observable A $\otimes$ B, the Generalized Born Rule applies. The probability that A $\otimes$ B takes a particular $a_i b_j$ value is given by:

$$\text{Prob}_{W1+2} (A \otimes B = a_i b_j) = \text{Tr} (W_{1+2} P_{ij})$$

and the expectation value of the "total" A $\otimes$ B observable in state W_{1+2} is:

$$\text{Exp}_{W1+2} (A \otimes B) = \text{Tr} ((A \otimes B) W_{1+2})$$

We will sometimes be given the state W_{1+2} of a composite system, and then asked to figure out what the reduced states W_1, W_2 of the separated sub-systems must be. Given a couple of observables A and B on H_1, H_2, there are some relatively straightforward identifications that help to work out the reduced states, namely:

$$\text{Tr} ((A \otimes I) W_{1+2}) = \text{Tr} (A W_1)$$

$$\text{Tr} ((I \otimes B) W_{1+2}) = \text{Tr} (B W_2)$$

where I is the identity observable. This amounts to the demand that the probability distribution over the eigenspaces of observable A (B) defined by the reduced state W_1 (W_2) be the same as that laid out over A $\otimes$ I (I $\otimes$ B) by the composite state W_{1+2}, thus effectively ensuring that the choice of description (either in the larger or smaller Hilbert space) of a subsystem in a larger composite system has no measurable consequences as regards the monadic properties of the individual subsystems.

APPENDIX 2: THE FORMAL CONDITIONS OF THE QUANTUM MEASUREMENT THEORY

To formally describe the problem of measurement, we need to first introduce some notation, and denote by Prob(W, Q) the probability distribution defined by $\text{Prob}_w(Q = q_n)$, for all eigenvalues q_n of Q. And let us denote *Q-indistinguishable* states W, W' as W ≡ W'. Two states W, W' are *Q-indistinguishable* if and only if Prob(W, Q) = Prob(W', Q).

We may now enunciate the following two conditions on measurement interactions:

The Transfer of Probability Condition (TPC)

$$\text{Prob}\left(U(W_o \otimes W_a)U^{-1}, I \otimes A\right) = \text{Prob}\left(W_o, O\right)$$

This condition expresses the requirement that the probability distribution over the possible outcomes of the relevant observable O of the object system should be reproduced as the probability distribution over possible outcomes of the pointer position observable in the final state of the composite (object + apparatus) system. (TPC) entails the following minimal condition on measurements employed by Fine (1970) and Brown (1985): A unitary interaction on a (object + apparatus) composite is a W_a measurement only if, provided that the initial apparatus state is W_a, any two initial states of the object system that are O-distinguishable are taken into corresponding final states of the composite that are (I ⊗A)- distinguishable. So we can use the pointer position of the measuring apparatus to tell apart two initial states of the object system that differ with respect to the relevant property.

The Occurrence of Outcomes Condition (OOC)

$$U(W_o \otimes W_a)U^{-1} = \Sigma\, c_n\, W_n \text{ where } \forall W_n\, \exists\, \mu_n\colon \text{Prob}_{W_n}(I \otimes A = \mu_n) = 1$$

This condition is often taken to express the requirement, inspired by the eigenstate–eigenvalue link, that the final state of the composite be a mixture over eigenstates of the pointer position observable. But to be precise, it expresses the more general idea that the final state of the composite must be a mixture over states in each of which the pointer position observable takes one particular value or other with probability 1.

NOTES

1. I am grateful for comments and reactions from audiences at the Fictions conference at Madrid's Complutense University (2006), the APA Pacific

Division conference in Oregon (2006), the ZiF workshop at Bielefeld (2007), the Dubrovnik IUC conference (2007), and the workshop on scientific models at the University of Barcelona (2007). Thanks in particular to my commentators Carl Hoefer, Alfred Nordmann, and Eric Winsberg, and to Ron Giere for his detailed comments on a draft version of the chapter.

2. Vaihinger (1911, 1924).
3. Vaihinger (1924, p. 99).
4. Fine (1993).
5. For some examples see the essays collected in the volumes edited by Morrison and Morgan (1999), Magnani et al. (1999), Jones and Cartwright (2004), and de Chadarevian and Hopwood (2004).
6. Vaihinger (1924, p. 97ff).
7. This agrees with much of the recent literature on fictions in metaphysics and aesthetics—which tends to distinguish carefully fiction from falsehood. For instance, in Walton's (1990) account of fiction as make believe, what characterizes fictions is their role as props in the prescription of imaginings, regardless of whether these turn out to be true or false. (Walton's theory strikes me as inappropriate for scientific fictions for other reasons, which are not relevant to my discussion here.)
8. Fine draws a similar distinction between virtuous and vicious fictions (Fine, 1993, p. 5).
9. In the account that follows I am indebted to Morrison (2000, esp. chap. 3) and Warwick (2003, esp. Chapters 6 and 7).
10. See also Morrison's extended discussion in this volume.
11. For instance, Trouton's and Larmor's ontological commitments to the ether could not have been more different. Trouton took second-order effects of the ether's motion with respect to the earth to be measurable by ordinary optical instruments on the earths' surface, whereas Larmor took any observable effects to be impossible as a matter of principle. See Warwick (1995).
12. Hesse (1966).
13. My exposition is indebted to the essays in Buchwald and Warwick (2004), and in particularly to Chapter 1, Smith (2004).
14. Geiger and Marsden (1909).
15. Rutherford (1911).
16. Thomson (1903, p. v).
17. Stellar astrophysics models provide a nice illustration of yet another thesis of mine, namely, that in scientific representation often the target is constructed along with the representation itself. I leave the detailed defense of this claim for another occasion (but see Suárez, 1999, for a preliminary account).
18. One of the earliest such proofs is due to Fine (1970). For details see Suárez (2004a).
19. Some of these options are described in Busch et al. (1991).
20. Gombrich (1984, Chapter 1).
21. I agree with Lopes (1996, p. 79) that this requirement is too strong on substitutive accounts, and would yield contradiction.
22. This "substitutive" character of models has been noted in the recent modeling literature, particularly in the "mediating models" literature (Morgan & Morrison, 1999), where models replace their targets as the main "focus of scientific research."
23. In *fictive* representation, the modeled entity (the target) is a real entity. In *fictional* representation, however, the target is also an imagined entity, just like the model.
24. Not all inferences will do, only those specific inferences that do not follow merely from the representational relation itself between A and B.

25. Suárez (2004b, p. 770).
26. For more on the notions of "constituents" and "means" see Suárez (2003, 2004b).
27. I am assuming here that it makes sense to speak of a *relation* between an existing model and a fictional entity or process; the assumption that some intensional relations might lack the corresponding extensions is of course controversial, but it does not alter anything substantial in what follows.
28. The names of Ron Giere and Bas Van Fraassen are often associated to such accounts, but I believe the association is not altogether fair. Both Van Fraassen (1994) and Giere (2004) make it clear that intentional elements are also needed for representation. But I am not here arguing against straw men. Others have embraced such accounts on their behalf, and many others have assumed that the accounts were properly theirs.
29. To speak of the "mathematical structure" of concrete objects, such as plum puddings, is problematic in any case, because there is a well-known issue of underdetermination of any abstract structure by a concrete object or sets of objects.

10 A Function for Fictions
Expanding the Scope of Science

Eric Winsberg

1. INTRODUCTION

To a first approximation, fictions are representations that don't concern them-selves with truth. Science, to be sure, is full of representations. But the repre-sentations offered to us by science, or so we are inclined to think, are supposed to aim at truth (or at least one of its cousins: approximate truth, empirical ade-quacy, reliability). If the proper and immediate object of fictions is contrary to the aims of science, what role could there be for fictions in science?

This chapter argues for at least one important role for fictions in science, especially in the computationally intensive sciences of complex physical sys-tems—in computer simulation. Fictions, I will argue, are sometimes needed for extending the useful scope of theories and model-building frameworks beyond the limits of their traditional domains of application. One especially interesting way in which they do this is by helping to enable model builders to sew together incompatible theories and apply them in contexts in which neither theory by itself will do the job.

2. FICTIONS VERSUS MODELS

I begin the chapter by clarifying what I take it to mean for a representation in science to be a fiction. My view will differ substantially from many oth-ers who discuss the role of fictions in science. The history of discussions of fictions in the philosophy of science goes back at least to Hans Vaihinger's famous book, *The Philosophy of 'As If.'* In Vaihinger's view, science is full of fictions.[1] This is a view that, I think, is shared by many who discuss fictions in science, including some of the contributors to this volume. In Vaihinger's view, any representation that contradicts reality is a fiction or, at least, a semi-fiction. (A full fiction, according to Vaihinger, is something that con-tradicts itself.) And so, in this view, most ordinary models in science are a kind of fiction.

"Fictional," however, is not the same thing as "inexact" or "not exactly truthful." Not everything, I would argue, that diverges from reality, or from our best accounts of reality, is a fiction. Many of the books to be found in the nonfiction section of your local bookstore will contain claims that are inexact, or even false. But we do not, in response, ask to have them reshelved in the fiction section. An article in this morning's newspaper might make a claim, "The green zone is a 10-km^2 circular area in the center of Baghad," which is best seen as an idealization. And though I live at the end of a T-intersection, "Google Maps" shows the adjacent street continuing on to run through my home. Still, none of these things are fictions.

I take as a starting point, therefore, the assumption that we ought to count as nonfictional many representations in science that fail to represent exactly: even representations that in fact contradict what our best science tells us to be the case about the world. Many of these kinds of representations are best captured by our ordinary use of the word "model." The frictionless plane, the simple pendulum, and the point particle all serve as good representations of real system for a wide variety of purposes. All of them, at the same time, fail to represent exactly the systems they purport to represent, or, for that matter, any known part of the world. They all incorporate false assumptions and idealizations.

But, contra Vaihinger, I urge that, because of their function (in ordinary contexts), we continue to call these sorts of representations "models," and resist calling them fictions. It would seem to me to be simply wrong to say that ordinary models in science are not concerned with truth or any of it cousins. In sum, we do not want to get carried away. We do not want all (or almost all) of the representations in science, on maps, in journalism, and so on to count as fictions. To do so risks not only giving a misleading overall picture of science ("all of science is fiction!") but also weakening to the point of emptiness a useful dichotomy between different kinds of representations—the fictional and the nonfictional. If only representations that are exact are nonfictions, then even the most ardent scientific realist will have to admit that there are precious few nonfictional representations in the world.

Fictions, then, are more rare in science than are models. That's because most models in science aim at truth or one of its cousins. But fictions do not. It might seem, then, that in this more narrow conception of what it is for a representation to be a fiction, there will probably turn out to be no fictions in science. Part of my goal, then, is to show that there are. To do this, I will have to define my more limited conception of what it is to be a fiction. This will involve being more clear about what it means to "aim at truth or one of its cousins."

So how should we proceed in demarcating the boundary between fictions and nonfictions? The truly salient difference between a fictional and a nonfictional representation, it seems to me, rests with the proper *function* of the representation. Indeed, I argue that would should count any representation—even one that misrepresents certain features of the world (as most models do)—as a nonfiction if we offer it *for* the sort of purpose for which we ordinarily offer nonfictional representations.

I offer, in other words, a pragmatic, rather than a correspondence, conception of fictionality. What, then, is the ordinary function of nonfictional representations? I would suggest that, under normal circumstances, when we offer a nonfictional representation, we offer it for the purpose of being a "good enough" guide to the way some part of the world is, for a particular purpose. Here, "good enough" implies that the model is accountable to the world, (in a way that fictions aren't), in the context of that purpose. On my account, to hold out a representation as a nonfiction is ipso facto to offer it *for* a particular purpose and to *promise* that for that purpose, the model "won't let you down," when it comes to offering guidance about the world for that purpose. In short, nonfictional representations promise to be *reliable* in a certain sort of way.

But not just in any way. Consider an obvious fiction: the fable of the grasshopper and the ant. The fable of the grasshopper and the ant, you may recall, is the story of the grasshopper who sings and dances all summer while the ant toils at collecting and storing food for the coming winter. When the winter comes, the ant is well prepared, and the grasshopper, who is about to starve, begs for his charity. This is what we might call a didactic fiction. The primary function of the fable, one can assume, is to offer us important lessons about the way the world is and how we should conduct ourselves in it. For the purpose of teaching children about the importance of hard work, planning and the dangers of living for today, it is a reasonably reliable guide to certain features of the world (or so one might think).

So why is it a fiction? What is the difference, just for example, between didactic fictions and nonfictions if both of them can serve the purpose of being reliable guides? To answer this, I think we need to consider the representational targets of representations. The fable of the grasshopper and the ant depicts a particular state of affairs. It depicts a land where grasshoppers sing and dance, insects talk, grasshoppers seek charity from ants, etc. If you read the fable wrong, if you read it as a nonfiction, you will think the fable is describing some part of the world—its representational target. And if you read it as a nonfiction, then you will think it is meant to be a reliable guide to the way this part of the world—this little bit of countryside where the grasshopper and the ant live—is. In short, the fable is a useful guide to the way the world is in some general sense, but it is not a guide to the way its prima facie representational target is. And that is what makes it, despite its didactic function, a fiction.

Nonfictions, in other words, aren't just reliable guides to the way the world is in any old way. They describe and *point to a certain part of the world* and say "if you want to know *about that part of the world I am pointing to,* for a certain sort of purpose, I promise to help you in that respect and not let you down." The importance of this point about the prima facie representation targets of representations will become clear later.

Fictional representations, on the other hand, are not thought to be good enough guides in this way. They are offered with no *promises of a broad domain of reliability.* Unlike most models in science, fictions don't come stamped with promissory notes that say something like "in these respects and to this degree of accuracy (those required for a particular purpose), some domain of the world—the domain that I purport to point to—is like me, or will behave like me."

So crucial to understanding the difference between fictional and nonfictional representations is to understand the different functions that they are intended to serve. But intended by whom? A brief note about intentionality is in order: It probably sounds, from the above, as though in my view whether or not a representation counts as a fiction depends on the intention of the author of the representation. On such a view, if the author *intends* for the representation to carry with it such a promissory note, then the representation is nonfictional. This is close to my view. But following Katherine Elgin (this volume), I prefer to distinguish fictions from nonfictions without reference to the intention of the author. If we find, someday, the secret diaries of James Watson and Francis Crick, and these reveal to us that they intended the double helix model as a planned staircase for a country estate, this has no bearing on the proper function of the model. In my view, what a representation is for depends not on the intention of the author, but on the community's norms of correct use.

Consider the famous tapestry hanging in the Cloisters museum in New York, the Unicorn in Captivity. This is a nice example of a representation. Quite possibly, it was the intention of the author that the tapestry be taken as a nonfictional representation belonging to natural history. More importantly, his contemporaries probably believed in unicorns. Had there been museums of natural history at the time, this might have been where the tapestry would have hung. But today, the tapestry clearly belongs in a museum of art. In deciding whether the tapestry is a fiction or a nonfiction does not depend on the intention of the author, it depends on what, based on the community's norms, we can correctly take its function to be—on what kind of museum the community is likely to display it in. Once upon a time, that role might have been in providing a guide to the animal kingdom. But it no longer is.

3. FICTIONS IN SCIENCE

It is clear that science is full of representations that are inexact, and of representations that contradict what we know or believe to be the case about the world. It is clear, that is, that science is full of models. But are there fictions in science? If a representation is not even meant to serve as a reliable guide to the way the world is, if it's a fiction, wouldn't it necessarily fall outside of the enterprise of science?

Despite the rather liberal constraints I want to impose on counting as a nonfiction, I believe there are a variety of roles that fictions can play in science. I want to outline one such role here. To do so, I turn now to two examples of computer simulations in the physical sciences where, I argue, fictions are employed. The first example comes from simulation methods in a field called "nanomechanics," and the second comes from computational fluid dynamics. In both of these examples, model builders add fictions to their models in order to extend the useful scope of the model-building frameworks they employ beyond the limits of their traditional domains of application.

The first example comes from a set of methods in nanomechanics called "parallel multiscale" simulations, (or sometimes "concurrent coupling of length scales").[2] These methods were developed by a group of researchers interested in studying the mechanical properties (how they react to stress, strain, and temperature) of intermediate-sized bits of solid-state silicon. What makes these modeling technique "multiscale" is that they seamlessly couple together the effects described by three different levels of description: quantum mechanics, molecular dynamics, and continuum mechanics. By "sewing" together these three model-building frameworks, model builders are able to go beyond the useful scope of any one of the three alone.

Modelers of nanoscale solids need to use these multiscale methods—the coupling together of different levels of description—because each individual theoretical framework is inadequate on its own at the scale in question. The traditional theoretical framework for studying the mechanical behavior of solids is continuum mechanics (CM). CM provides a good description of the mechanics of macroscopic solids close to equilibrium. But the theory breaks down under certain conditions. CM, particularly the flavor of CM that is most computationally tractable—linear elastic theory—is no good when the dynamics of the system are too far from equilibrium. This is because linear elastic theory assumes that materials are homogeneous even at the smallest scales, when in fact we know this is far from the truth. It is an idealization. When modeling large samples of material, this idealization works, because the sample is large enough that one can effectively average over the inhomogenaities. Linear elastic theory is in effect a statistical theory. But as we get below the micrometer scale, the fine-grained structure begins to matter more. When the solid of interest becomes smaller than approximately one micron in diameter, this "averaging" fails to be adequate. Small local variations from mean structure, such as material decohesions—an actual tearing of the material—and thermal fluctuations, begin to play a significant role in the system. In sum, CM cannot be the sole theoretical foundation of "nanomechanics"—it is inadequate for studying solids smaller than one micrometer in size (Rudd & Broughton, 2000).

The ideal theoretical framework for studying the dynamics of solids far from equilibrium is classical molecular dynamics (MD). This is the level at which thermal fluctuations and material decohesions are most naturally described. But computational issues constrain MD simulations to about

10^7–10^8 molecules. In linear dimensions, this corresponds to a constraint of only about 50 nanometers.

So MD methods are too computationally expensive, and CM methods are insufficiently accurate, for studying solids that are on the order of 1 micrometer in diameter. On the other hand, parts of the solid in which the far-from-equilibrium dynamics take place are usually confined to regions small enough for MD methods. So the idea behind multiscale methods is that a division of labor might be possible—use MD to model the regions where the real action is, and use CM for the surrounding regions, where things remain close enough to equilibrium for CM to be effective.

There is a further complication. When cracks propagate through a solid, the process involves the breaking of chemical bonds. But the breaking of bonds involves the fundamental electronic structure of atomic interaction. So methods from MD (which use a classical model of the energetic interaction between atoms) are unreliable right near the tip of a propagating crack. Building a good model of bond breaking in crack propagation requires a quantum mechanical (QM) approach. Of course, QM modeling methods are orders of magnitude more computationally expensive than MD. In practice, these modeling methods cannot model more than 250 atoms at a time.

The upshot is that it takes three separate theoretical frameworks to model the mechanics of crack propagation in solid structures on the order of one micron in size. Multiscale models couple together the three theories by dividing the material to be simulated into three roughly concentric spatial regions. At the center is a very small region of atoms surrounding a crack tip, modeled by the methods of computational QM. In this region, bonds are broken and distorted as the crack tip propagates through the solid. Surrounding this small region is a larger region of atoms modeled by classical MD. In that region, material dislocations evolve and move, and thermal fluctuations play an important role in the dynamics. The far-from-equilibrium dynamics of the MD region is driven by the energetics of the breaking bonds in the inner region. In the outer region, elastic energy is dissipated smoothly and close to equilibrium, on length scales that are well modeled by the linear-elastic, continuum dynamical domain. In turn, it is the stresses and strains applied on the longest scales that drive the propagation of the cracks on the shortest scales.

Researchers describe theses sorts of phenomena as "inherently multiscale." This means that what is required for simulating these phenomena is an approach that simulates each region simultaneously, at its appropriate level of description, and then allows each modeling domain to continuously pass relevant information back and forth between regions—in effect, a model that seamlessly combines all three theoretical approaches. What allows the integration of the three theories to be seamless is that they overlap at the boundary between the pairs of regions. These boundary regions are where the different regions "shake hands" with each other. The regions are called the "handshaking regions" and they are governed by "handshaking algorithms."

The basic idea behind the handshaking algorithms is rather simple in principle, but becomes rather complicated in practice. Each of the three basic regions has its own computational scheme based on a function that gives the energy of interaction between the simplest 'elements' of the scheme. In the CM region, the elements are the vertices of the grid over which the continuous equations have been discretized. The energy function comes from the elastic forces. In the MD region, the elements are molecules, and the energy function comes from a classical force function. And in the QM region, the elements are atoms, and the energy function is a quantum Hamiltonian. What the handshaking algorithms do is to match up the elements from the neighboring regions and when calculating the energetic interaction of, say, a vertice and an atom, to take the average of the value that each of the two different schemes provide.

Suppose, for example, that there is an atom on the MD side of the border. It looks over the border and sees a grid point. For the purpose of the handshaking algorithm, we treat that mesh point as an atom, calculate the energetic interaction according to the classical potential, and we divide it by two (remember, we are going to be averaging together the two energetics.) We do this for every atom/mesh point pair that spans the border. This is one half of the "handshaking Hamiltonian." The other half comes from the continuum dynamics' energetics. Whenever a mesh point on the CM side of the border looks over and sees an atom, it pretends that atom is a mesh point. Thus, from that imaginary point of view, there are complete tetrahedra that span the border (some of whose vertices are mesh points that are "really" atoms.) Treating the position of that atom as a mesh point position, the algorithm can calculate the strain in that tetrahedron, and integrate over the energy stored in the tetrahedron. Again, because we are averaging together two Hamiltonians, we divide that energy by two.

The general approach for the handshaking algorithm between the quantum region and the molecular dynamics region is similar: The idea is to create a single Hamiltonian that seamlessly spans the union of the two regions. But in this case, there is an added complication. The difficulty is that the algorithm used to calculate the energy in the QM region does not calculate the energy locally. That is, it doesn't apportion a value for the energy for each interatomic bond; it calculates energy on a global basis. Thus, there is no straightforward way for the handshaking algorithm between the quantum and MD region to calculate an isolated quantum mechanical value for the energetic interaction between an outermost quantum atom and a neighboring innermost MD atom. But it needs to do this in order to average it with the MD value for that energy.

The solution that researchers have developed to this problem is to employ a trick that allows the algorithm to localize that QM value for the energy. The trick is to employ the convention that at the edge of the QM region, each "dangling bond" is "tied off" with an artificial univalent atom. To do this, each atom location that lies at the edge of the QM region is assigned an atom with a hybrid set of electronic properties. In the case of silicon, what is needed is something like a silicon atom with one valence electron. These atoms, called "silogens," have some of the properties of silicon and some of the properties of hydrogen. They produce a bonding energy with other silicon atoms that is equal to the usual Si–Si bond energy, but they are univalent like a hydrogen atom. This is made possible by the fact that the method is semi-empirical, so values for matrix elements can simply be assigned at will and put in "by hand." This makes it such that the silogen atoms don't energetically interact with their silogen neighbors, which means that the algorithm can localize their quantum mechanical energetic contributions. Finally, once the problem of localization is solved, the algorithm can assign an energy between atoms that span the threshold between regions that is the average of the Stillinger–Weber potential and the energy from the Hamiltonian in the tight-binding approximation. Again, this creates a seamless expression for energy.

Silogen atoms, I want to argue, are fictions. To see this, we need to look at their function. But we need to be careful. If we view the overall model that drives the simulation as a whole, it is clearly nonfictional. The representational target of the Abraham model is micrometer-sized pieces of silicon. And the Abraham model is meant to be reliable guide to the way that such pieces of silicon behave. To endorse such a model in the relevant respect is *to promise that* for the purpose of designing nano-electromechanical systems (NEMS), *the model will not let you down*. It will be a good enough guide to the way these pieces of silicon behave—accurate to the degree, and in the respects, necessary for NEMS design. Though it contradicts reality in more ways than we could probably even count, Abraham's model is a nonfiction.

But within the simulation, we can identify components of the model that, prima facie, play their own local representational role. Each point in the QM region appears to represent an individual atom. Each point in the MD region, and each tetrahedron in the FE region, has an identifiable representational target. Some of these points, however—the ones that live in the QM/MD handshaking region—are special. These points, which represent their representation targets *as* silogen atoms, do not function to be reliable guides to the way the way those atoms behave. Silogens are *for* tying off the energy function at the edge of the QM region. They are *for* getting the whole simulation to work properly, not for depicting the behavior of the particular atom being modeled. We are deliberately getting things wrong locally so that we get things right globally. The silogen atoms are fictional entities that "smooth over" the inconsistencies between the different model-building frameworks, and extend their scope to domains where they would individually otherwise fail. In a very loose sense, we can think of them as being similar to the fable of the grasshopper and the ant.

In the overall scheme of things, their grand purpose is to inform us about the world. But if we read off from them their prima facie representational targets, we are pointed to a particular domain of the world about which they make no promises to be reliable guides.

The second example comes from a method in computational fluid dynamics (CFD) used to model the supersonic flow of fluids with strong shocks.[3] Shock waves are notoriously difficult to simulate numerically. The reason is that the computational models involved use spatial "grids," and an abrupt shock will fall between the grid lines. This, in turn, makes the energy of the simulation blow up. John von Neumann came up with the idea of trying to blur the shock over a few grid cells. He did this by making up an "artificial viscosity" for the fluid. The trick is to treat the fluid as extremely viscous right where the shocks are. This makes the shocks thicker than a few grid cells, and the simulation will no longer blow up.

Many simulations in CFD employ methods of discretization. Discretization begins with a set of mathematical equations that depict the time evolution of the system being studied in terms of rules of evolution for the variables of the model. In the case of CFD, the equations in question are some version of either the Euler equations or the Navier–Stokes equations, depending on the factors to be included, and the coordinate system to be employed. To study a particular flow problem, it would be desirable to find solutions to those models' equations. With most of the systems we encounter in fluid dynamics, however, the models that are suggested directly by theory consist of second-order, nonlinear differential equations. It is often impossible, even in principle, to find closed-form solutions to these equations.

What a simulationists must do, therefore, is to find a replacement for the model hatched out of the theory—one that that consists of a set of equations that can be iterated numerically, using step by step methods. When a theoretically motivated model is thus "discretized," and turned into a simulation model, the original differential equations are transformed into difference equations, and crafted into a computable algorithm that can be run on a digital computer. The finite "difference" between the values of spatial and temporal variables is sometimes called the "grid size" of the simulation, because one can imagine a spatiotemporal grid on which the simulation lives.

One of the earliest uses of finite difference simulations arose in connection with the Manhattan Project during World War II. John von Neumann and his group of researchers used finite difference computations to study the propagation and interaction of shock waves in a fluid, a subject crucial to the success of the atomic bomb.

We generally think of shock waves as abrupt discontinuities in one of the variables describing the fluid, but it was quickly recognized that treating them in this way would cause problems for any numerical solution. The reason is that a shock wave is not a true physical discontinuity, but a very narrow transition zone whose thickness is on the order of a few molecular mean free paths. Even with today's high-speed and high-memory computers, calculating fluid flow with a differencing scheme that is fine enough to resolve this narrow transition zone is wildly impractical. On the other hand, it is well known that a simulation of supersonic fluid flow that does not deal with this problem will develop unphysical and unstable oscillations in the flow around the shocks. These oscillations occur because of the inability of the basic computational method to deal with the discontinuities associated with a shock wave—the higher the shock speed, the greater the amplitude of these oscillations becomes. At very high speeds, such a simulation quickly becomes useless. To make it more useful and accurate, what simulationists need to do is to somehow dampen out these oscillations.

The generally accepted way to do this, which was originally devised by von Neumann and Richtmyer while working at Los Alamos, is to introduce into the simulation a new term, an "unphysically large value of viscosity," which is called "artificial viscosity." The inclusion of this term in the simulation is designed to widen the shock front and blur the discontinuity over a thickness of two or three grid zones. This enables the computational model to calculate certain crucial effects that would otherwise be lost inside one grid cell: in particular, the dissipatcion of kinetic energy into heat.

A standard equation for the value of artificial viscosity at a particular point in a simulation is given here:

$$Q = \{ \begin{array}{ll} l^2(\Delta x)^2 \rho (\frac{\partial v}{\partial x})^2 & \frac{\partial v}{\partial x} < 0 \\ 0 & \text{otherwise} \end{array}$$

where l is a dimensionless constant, delta x is the grid size, rho is the density, and v is the local value of the velocity. This makes the value of viscosity proportional to the local vorticity of the fluid, and guarantees that it will be extremely high only in the local presence of a shock.

The high viscosity attributed to the fluid near the shocks is unrealistic. It contradicts our best accounts of what the world is actually like at the level of description in which fluids operate. In some respects, it is like the frictionlessness of an inclined plane, in that, just as it is literally false of any actual plane in the world that it is frictionless, it is literally false of any fluid that it is has a viscosity that varies in proportion to its local vorticity. In that respect, von Neumann–Richtmyer models are not fictions. They are just models. They are representations that contain substantial inaccuracies, but that for certain purposes (including the purpose, unfortunately, of designing thermonuclear weapons) are "good enough" guides to the way their representational targets (the entire fluid system in question) behaves.

But models of fluids with shocks are unusual creatures in certain respects. Ever since Ludwig Prandtl introduced the idea of an aerodynamic boundary layer, where the flow of a fluid is divided into two flow fields—one inside the layer, where viscosity matters most, and one in the rest of the fluid, where viscosity can be ignored—fluid dynamicists have become accustomed to the idea that models of fluids are compartmentalized. And it is traditional to think of shocks in the same way. The model of a fluid with shocks is composed of a submodel of the shock, and a submodel of the rest of the fluid.

If we think of models of fluids in this way, then the model of the shock inside the von Neumann–Richtmyer (vNR) model is a fiction. No one thinks: "Treat the fluid inside a shock as having an enormous value for viscosity and you won't go wrong in understanding the behavior of the shock." In fact, if you look at the behavior of the fluid in the regions where you have applied the high viscosity, you are getting things all wrong—deliberately so: You have virtually eliminated the existence of abrupt shocks. This is not a formula for gaining understanding of shocks. It is a formula for allowing the overall simulation to survive the computational disasters that shocks usually bring about. The hope is that that's OK. The hope is that by getting things dramatically wrong near where the shocks will be, you will prevent energy blowups in your algorithm, and you will get things reasonably right overall. If we think of the vNR method as something that we use to build models of fluids, it is a nonfiction, and artificial viscosity is a nonfictional property of the fluid. But if we think of the *models of shocks* that we build with the vNR method, these are fictions, and the high viscosity is a fictional characteristic of the fluid in the shock.

So artificial viscosity isn't *for* being a reliable guide (even in certain limited respects and to some limited degree of accuracy) to the way fluids in shocks behave. What silogen atoms and artificial viscosity have in common (though one is an entity, the other a property) is just that: They both point at some part of the world, but neither is for getting the behavior of that part of the world right. They add to the overall reliability of the simulation by sacrificing the reliability of the simulation in getting things right about the part that they represent.

Despite the apparent conflict of the function of fictions with the aims of science, fictions, it would seem, play at least one important role in science. They help us to sew together inconsistent model-building frameworks, and to extend those frameworks beyond their traditional limits.

NOTES

1. My understanding of Vaihinger comes almost entirely from Fine (1993).
2. My discussion of parallel multiscale simulations draws from several sources. Included are Abraham et al. (1998), Broughton et al. (1999), Chelikowski and Ratner (2001), Nakano et al. (2001), and Rudd and Broughton (2000). For more details, see Winsberg (2006b).
3. My discussion of parallel multiscale simulations draws from the following sources: Campbell (2000), von Neumann and Richtmyer (1950), and Caramana et. al. (1988) For more details, see Winsberg (2006a).

Part V
Fictions in the Special Sciences

11 Model Organisms as Fictions

Rachel A. Ankeny

... fiction is fact distilled into truth.[1]

1. INTRODUCTION

Conventional (albeit oversimplified) wisdom about science holds that it involves a search for truth or for an accurate reflection of the way the world is. However, the biological sciences present us with a much more complex picture of the natural world, one riddled by variation and change. Leaving aside philosophical debates over the uniqueness of biology and the possibility of its unification with or reduction to other sciences, its lack of laws, and whether fundamental processes and mechanisms can be identified that underlie what may merely be the appearance of diversity, it is indisputable that many actual practices in contemporary biological sciences, particularly those associated with genomic techniques, explicitly focus on elucidating the shared or fundamental components of various natural processes and entities. In the process, assorted methods are utilized to impose what at times may be a somewhat selective order, so as to be able to control this variation and hence more efficiently study and understand it.

This essay examines one such set of practices in the biological sciences that are associated with an early phase in a particular field of research, namely, the development of a range of biological resources associated with a particular subset of experimental organisms, the so-called 'model organisms.' I illustrate how these resources rely on the creation of what can be viewed as explicitly fictitious entities, and which often are recognized as such by the scientists who work with them. The creation of these fictions in turn produces an epistemic space that would not be accessible without such fictions, and serves as a springboard to further research and refinement of scientific understanding, particularly the future development of explanations and theories as well as biotechnological products. Understanding these biological entities as a form of fictions can provide insights into the ontological and epistemological assumptions that ground their use for research, as well as their limitations and prospects.

2. A BRIEF NOTE ABOUT FICTION

There is an extensive literature about how literary works, especially fictional ones, can be philosophical and particularly how they might be able to provide moral guidance or conceptual understanding.[2] I do not wish to develop an argument here in favor of a particular view on the role of fiction, but selectively rely on discussions regarding its potential role in fostering conceptual understanding. I believe these are most relevant to considerations about the use of fictions in science and hence draw on them in order to illustrate the possible utility of such approaches without conclusively defending them. For instance, Eileen John has argued that fictions are not philosophical works in themselves, but that our responses to a work of fiction can involve calling us to the pursuit of conceptual knowledge, as fiction (like science) is a knowledge-gathering enterprise: "works of fiction can also be forces of articulation, bringing specific conceptual questions to the surface. The characters and events of the story can pose well-defined questions, in response to which we can think concretely about what difference it would make to use a concept one way or another."[3]

David Novitz suggests that fiction allows us to acquire conceptual or cognitive skills that "offer radically new ways of thinking about or perceiving aspects of our environment."[4] Hence from fictions we can learn about and achieve a deeper understanding of the actual world around us, not only about what is contained in it and how these things function, but how they might work in novel situations and how we can form generalizations from specific instances. This type of understanding may occur through a range of mechanisms, for example, by allowing us to notice associations, to compare the new data to our existing accumulated empirical knowledge, to consider and assess our evaluative standards, and to reveal the networks of concepts that we typically use and revise or abandon them as necessary to achieve a better fit with the actual world, while nonetheless doing "descriptive justice to a particular case," as John notes.[5]

Research in the biological sciences, particularly with model organisms, requires close attention to what 'descriptive justice' might in fact entail, and in this sense, the processes of construction and use of fictitious biological entities seem to share a number of attributes with the trade-offs in which authors engage when writing a work of historical fiction. We might intuitively think that history, like science, seeks to establish the truth, while fiction is invented, whereas in most cases the difference is not one of kind but of degree, inasmuch as historical work contains assumptions, postulations, or even inventions, and fictional work often relies on historical facts. It is the case that the two genres 'traffic in different kinds of truth,' as Matt Oja argues, with works at the historical end of the spectrum tending toward literal truth whereas those that are good examples of historical fiction rely on a mixture of literal and artistic truth.[6] As he notes, a purely fictional narrative would not have objective reality without being created, whereas fictional historical narratives draw on the real world at the same time as not being absolutely tied to it. There are a variety of reasons for this: The author may not have full information about the historical events in question or it might not serve his or her goals to stick too closely to what is known, be those goals creating dramatic effect, developing characters or plot, or even fostering conceptual understanding. In a similar spirit, working creatively with and around what is known, and even knowingly creating fictitious entities, in order to achieve a deeper understanding of natural phenomena, is an important part of scientific practice, as I show in a series of examples from model organism work in biology.

3. BACKGROUND: WHAT IS A MODEL ORGANISM?

The term 'model organism' is typically used to describe relatively simple or lower level organisms selected for examination because of the presumed applicability of data from such organisms more generally across the animal kingdom or at least to other, closely related organisms. These simpler organisms often are used as proxies for the actual objects of interest, such as more complex or higher level organisms including humans which are much more complex and difficult to study directly. The use of such organisms permits articulation of fundamental processes thought to be generally at work in many (or all) living things. Model organisms gained special prominence during Human Genome Project, which included complete mapping and sequencing of a number of these organisms as a means of refining technologies and gathering sequence data, key parts of which were predicted or known to be evolutionarily conserved and therefore shared across a range of organisms.[7]

Key examples of popular current-day model organisms include the yeast *Saccharomyces cerevisiae*, the bacterium *Escherichia coli*, the fruitfly *Drosophila melanogaster*, the nematode worm *Caenorhabditis elegans*, and various mouse strains.[8] Using model organisms as material for research takes a number of different forms, all of which require various simplifying assumptions to what is provided by the natural world. First, a standardized, laboratory wild type must be established based on a sampling of organisms that are thought to share (roughly) the same genotype and phenotype. Sometimes the choice of a particular wild type is rather arbitrary, and often reflects which strain of a particular organism is easiest to gather, maintain, and manipulate experimentally, rather than which strain or organismal type is most common or representative. Next, various resources are constructed from this wild type, to serve as what I have called elsewhere 'descriptive models' and capture details about various processes in the organism's life cycle; these include wiring diagrams (which provide complete information on the neural connections for a normal organism), charts of genetic sequence data, and cell lineage maps (which detail the various cell divisions from the one-cell stage to the complete organism).[9]

So-called 'digital organisms' are constructed to facilitate a more detailed understanding of the various processes associated with their generation and development. These resources typically are based on the wiring diagrams, genetic sequence data, and cell lineage maps, all of which begin as static resources, and permit reconstructions of various types of phenomena via computer simulation, such as visual simulations of the connections within the nervous system and of the stages of embryogenesis (the processes by which the embryo is formed and develops), including three-dimensional simulations of determination of cell position. Further, attempts are being made to construct novel synthetic organisms based on some simpler organisms; a key goal with these projects is to determine what is biologically essential to make a particular organism. Hence minimal, novel organisms are constructed that contain only the genes that perform these essential functions (and that do not have any extraneous or 'junk DNA'), as an interim step toward constructing fully synthetic organisms.

As no doubt is already apparent, each of these types of model organism research relies on a series of simplifying assumptions and choices that both impose control on any variations that might be detected and in some cases create distance between the object of research and what can actually be found in nature, as all of these resources are in some sense synthetic or at least selected from among the available natural variants. But are they a form of fictions, and more importantly, does viewing them as fictitious entities help us to better understand research practices in the modern biological sciences? In order to answer these questions, I provide a brief overview of each of these research practices, which stresses the features most relevant to consider their putative fictitious status, beginning with those that are the most overtly synthetic or artificial, in order to make the examples more compelling for the reader.

4. NOVEL SYNTHETIC ORGANISMS

The relatively new field of synthetic biology has as one of its ultimate goals the artificial production of living cells and ultimately organisms, using chemicals or other synthetic means, which could be engineered for specific purposes such as making more efficient or cost-effective biotechnological products. Preliminary results from recently published research indicate success in rebuilding in the laboratory the genome of a bacterium that lives parasitically within the human urogenital tract, *Mycoplasma genitalium*.[10] *M. genitalium* has been a central organism for this type of research because its minimal metabolism and relatively nonredundant genome make it a good approximation of the minimal set of genes necessary for bacterial life forms.

The first necessary step for creation of novel synthetic organisms is the determination of which parts of the naturally existing organisms are absolutely necessary, in other words, the isolation of the smallest set of genes (the 'minimal genome') that allows for survival and replication of an organism in a certain environment. So for instance, following sequencing of the genome of *M. genitalium,* which is the smallest organism that can be grown in pure (axenic) culture and only has a single chromosome, using shotgun (whole-genome random) sequencing and assembly, it was initially proposed that only about 55% of protein-coding genes were essential under laboratory growth conditions.[11] This prediction was made using a theoretical (nonexperimental) comparison of its genome to those of other closely related bacteria and an assumption that those genes that are shared in common (and hence conserved evolutionarily) are essential. Other laboratories use computer modeling to determine the minimal number of biochemical pathways needed for basic metabolic and reproductive functions (see the next section on simulations for more details about these types of procedures).[12] Later research combining experimental and theoretical techniques showed that 382 of 482 genes were essential (with five other sets of disrupted protein-coding genes having apparently redundant essential functions).[13]

The second step toward the actual creation of a novel synthetic organism is reconstructing the minimal gene set, and then determining and providing the other conditions (i.e., any nongenetic ingredients) necessary for successful gene expression. This can be done in two ways: First, beginning from the intact natural organism, the set of genes thought to be unnecessary for basic functioning can be removed or inactivated. Alternatively, the proposed minimal genome can be synthesized and placed in environmental conditions that permit metabolic activity, survival, and replication. The latter technique was used in the recent construction of a synthetic version of *M. genitalium,* with short pieces of synthetic DNA that matched overlapping bits of the organism's genomic sequence linked together with enzymes and then combined into one complete chromosome. Except for nonessential markers ('watermarks') used to distinguish the artificial sequence from a natural one, the sequence produced synthetically was identical to the natural sequence of *M. genitalium.* Whether the sequence behaves and functions as a natural sequence is yet to be determined.

In summary, among the goals of these types of synthetic biological projects is to be able to reproduce natural processes in an artificial manner, not only for practical reasons but also in order to achieve a deeper understanding of how these fundamental processes work. As a journalist summarizes one of the key researcher's views: "The ultimate test of understanding a simple cell, more than being able to build one, would be to build a computer model of the cell, because that really requires understanding at a deeper level."[14] In order to truly understand nature, we must reconstruct it artificially.

5. DIGITAL ORGANISMS

Digital organisms rely on creating 'reconstructions' of phenomena of interest via computer simulation.[15] In the case of simulated model organisms, these include systems that allow visualization of the processes involved in embryogenesis, simulate neural connections, or allow three-dimensional mappings of cell position. The overall aim of the researchers who create these simulations is to obtain an understanding of the dynamics of a particular organism, which would not be possible (or at least would be extremely difficult) by working with the actual natural components and their functions under the usual experimental conditions, in part because such simulations explicitly ignore many known details about the nervous system.[16] The underlying claim is that an in-depth understanding may only be achievable through this type of reconstruction via modeling and simulation, as many of the phenomena of interest are complex and nonlinear. Due to complexity of biological entities and the systems that compose them, this simulated approach is necessary at least as a starting point for continuing research with the actual organisms themselves.

To take one example, in the nematode worm *C. elegans,* various neural circuits have been simulated and the neural networks trained using various computer algorithms in order to reproduce the neural behaviors in natural organisms. Lesion studies are then done on the simulation, where a portion of the network is ablated or destroyed to compare the resulting functioning (or loss thereof) in the simulation with data collected in real organisms. In turn, these results suggest hypotheses about the functions of various circuits and even single neurons that can be tested in experiments with real organisms.[17] Work using similar techniques has been used to trace learning behaviors, and thermotaxic and chemotaxic responses, among other complex phenomenon. There also have been attempts to simulate the full organism, and create a 'perfect' simulated organism, which would allow researchers access to a comprehensive visual database of the organism, and which in turn might allow a more general simulator to be developed for non-organism-specific morphogenesis.[18]

Partial or full digital organisms are typically constructed to achieve four main goals. First, their creators wish to create a space for hypothesis testing, for example, to bring genetic interactions (occurring across the genetic sequence) 'into focus' to see whether hypothetical genetic interactions can consistently recreate phenotypic changes that match those previously observed. Second, digital organisms allow predictions to be made; for instance, given a particular phenotypic phenomenon, a prediction can be made about which genotype is likely to produce it and the simulation can be checked to confirm this. Digital organisms also provide testable hypotheses, as they allow the design of experiments in the simulated environment and identify which mechanisms exist, in order to project them back onto the actual natural organism. Finally, digital organisms allow researchers to conceptualize global and local interactions (which might otherwise be overlooked in an actual organism) in order to identify and track what might be seen in 'real' experiments with actual organisms.

What are the criteria for what makes a useful digital organism or organismal system? First, the choice of the right level for abstraction is essential; for instance, complex phenotypic effects are often best understood when broken down into particular functions. In addition, trade-offs must occur between the sophistication of the simulation and its ability to recreate phenomena, particularly interactions between various systems or levels within the organism. The simulation system also should allow data to be replaced and other organisms thus to be modeled, thus fulfilling some basic criteria for generalizability. Those who develop digital organisms use an analogy to particle physics: The prediction of hypothetical particles at certain energy levels allows them to be detected, and similarly, digital organisms allow predictions to be made that can then be tested in the actual, live organisms.

Even crude simulations are useful. They allow understanding of complex phenomena and the development of a more 'intuitive understanding' of the overall processes, which is not possible by using (actual) organisms. However, the simulations have limits written into them. For instance, a cell position system works from existing data and makes various simplifying assumptions for example about direction of cell divisions, the locations of cells (in three dimensions), the estimated vertical position or depth of cells, and so on. Even with these limitations, such synthetic models allow testing of hypotheses, particularly where these hypotheses do not rely on the intricate details of each component that composes the system but more generally on their configurations.

In summary, as some of the advocates of these simulation approaches put it, "Realistic models require a substantial empirical database; it is all too easy to make a complex model fit a limited subset of data. Simplifying models are essential but are also dangerously seductive; a model can become an end in itself and lose touch with nature. Ideally these two types of models should complement each another."[19]

6. DESCRIPTIVE RESOURCES OF MODEL ORGANISMS

A central part of research with model organisms on which the previous approaches depend is the construction of a series of descriptive resources that capture details about a variety of processes and systems in the organism.[20] These include the wiring diagrams of neural connections, genomic sequence data, and cell lineage maps. In order to generate such resources, researchers make choices from among what may in some cases be variations even in normal organisms by making judgments about what is the standard or norm for the organism and what represents mere 'individual differences.' These resources are necessary for future hypothesis testing

and to proceed to building theories about the functions of various components, as well as understanding deviations from the norm such as production of abnormal phenotypes.

The goal of developing these descriptive models is to capture the most
commonly occurring structures or processes, by using multiple individual
specimens and creating a 'mosaic' organism. The wiring diagrams thus represent the 'consensus state' of an organism in terms of its neural patterns
and connections (and similar arguments can be made about the genetic
sequence data and the cell lineage diagrams). As the researchers themselves
put it, "we are reasonably confident that the structure . . . is substantially
correct and gives a reasonable picture of the organization of the nervous
system in a typical [organism]."[21]

7. EXPERIMENTAL WILD-TYPE ORGANISMS

In order for any research to occur with model organisms, it is necessary first
to establish a laboratory wild type, based on a sampling of various strains
that are thought to share (roughly) the same genotype and phenotype.
As has been noted in many case studies of the history of model organism
choice, prospective model organisms typically are selected and constructed
based not mainly on evidence about the universality or even typicality of
their biological characteristics and processes (though it is hoped that many
features will prove to be shared or common to other organisms), but primarily due to perceived experimental manipulability and tractability. For
example, *C. elegans* was chosen specifically for its developmental invari-
ance and simplicity, despite the atypicality of these biological characteristics (among many others) of *C. elegans,* even in comparison to other closely
related organisms.[22] The general aim of the original research project was
to achieve an understanding of developmental processes in metazoans (animals with bodies composed of differentiated cells, as opposed to protozoa
or unicellular animals).[23]

Oftentimes the choice of a particular wild type is rather arbitrary, and
reflects which strain of a particular organism is easiest to gather, maintain,
and manipulate experimentally, rather than which strain or organismal
type is most common in nature or representative in some other way. A key
part of any model organism research program is the selection of a wild
type and 'taming' of it, through processes of standardization, into a laboratory organism. In areas of biology where there is ongoing flow over time
between the laboratory and the field or the wild, there may be more than
one strain or variant that is held as a norm and thus more than one strain
or wild type might be used; however, particularly with model organisms
that are selected primarily because of their power for genetic analysis, it is
essential to settle on (and persist in using) one wild type, however arbitrary
it might be. Hence model organisms, particularly successful ones, are often

extremely distant from those that could be easily found outside the laboratory and in nature, which might be considered by some to be the 'real' organisms. However, this also means that there often are clear limits that are built into the experimental systems based on model organisms and that often are noneliminable, hence restricting researchers from making direct inferences back to the naturally occurring organisms.[24]

8. MODEL ORGANISM RESEARCH AND FICTIONS

As illustrated in the previous sections outlining various research programs based on model organisms, 'useful fictions' appear to be created, assumed, and endorsed in a variety of ways. Novel synthetic organisms perhaps provide the most obvious example. These organisms are explicitly built to mimic what occurs in nature but in a more streamlined manner. Although these extraneous components are naturally occurring and have developed historically in the organism over time, the goal of the research is to eliminate that which is unnecessary in favor of the minimal components that are required to create a functional and replicating organism. Hence these organisms represent a type of fictitious entity, similar to a narrative that might be based on real-life historical incidents but that strips away the unnecessary details in favor of retaining the overall form and intent of the story. These organisms are not only produced due to practical reasons (such as control of complexity), but also because eliminating the complications and details of what is real in favor of the fictitious allows revelation of the fundamental processes at work. The process of creating synthetic organisms in turn permits projection of the information obtained into novel situations, such as to permit the creation of a truly novel organism that is not naturally occurring but that represents the minimally necessary components for life, one of the long-term goals of some working in synthetic biology.

Digital organisms, which rely on computer simulation techniques, similarly allow elimination of complexities through explicitly ignoring details about the processes and functioning in the natural organism. In this sense, they are fictitious at least in comparison to the real model organisms from which they are derived, let alone those organisms that are to be actually found in nature. As with some types of fiction, they allow for clearer articulation of the relevant conceptual questions, and for the identification and testing of various pathways to see what changes minor modifications might make to the underlying system. Even crude simulations, in parallel to caricatures, can prove useful: They permit understanding of complex phenomena and particularly can allow a more intuitive understanding of the overall processes that is not possible by using actual messy and overly complicated natural organisms.

Even the descriptive resources that form an essential part of model organism research rely on the assumption of certain fictions, which include the elimination of variability in favor of canonical standardizations. These resources in a sense become a stock set of characters who generally behave in controllable and predictable ways, even if in reality (almost) no one would behave that way. In the process of people using them for research these resources serve as a jumping-off point for investigating what happens when we compare these to other instances such as abnormal specimens or closely related but different organisms. They allow reflection back to the real world and hence the potential for a deeper understanding of it.

Finally, wild-type model organisms themselves can be viewed as fictitious entities, again at least in comparison to what occurs in nature. They are carefully constructed with certain goals in mind for both the researcher who makes the initial choices and those who might use the organism in the future. Many of these choices might indeed be in a sense arbitrary, but the aim is to produce a coherent whole that has instrumental uses, including the ability to create a means for pursuing further testing and generation of knowledge. The story must hang together and achieve its goals. Although the high degrees of variability in the real world are well-recognized, these are traded off against the desire to pursue understanding fundamental or shared processes, mechanisms, and so on, those that might tell us something more generally about the nature of living things.

9. CONCLUSIONS: WHAT ARE FICTIONS GOOD FOR?

> Though fictions tell us things differently from the way they are, as if they were lying, fictions actually oblige us to examine them more closely, and they make us say: Ah, this is just how things are, and I didn't know it.[25]

Viewing various aspects of model organism research as different forms of fictions provides a helpful analogy for developing a better understanding of the epistemic potential and limits of research in this domain. First, in this type of research, the intention of the researchers is ultimately to investigate what actually takes place in nature; that is, these are not generally experiments in artificial life. This parallels what is seen by some as the goal of some forms of fiction: The aim is to cause us to reflect on similar phenomena in our experiences or in the world. Putting these points across in the form of a story allows a better understanding than would simple provision of a complete description of the actual situation.

Second, each type of model organism research outlined earlier involves certain simplifying assumptions and idealizations, which allow the creation of epistemic space which otherwise would not be accessible to researchers. If research were always done on the actual, naturally occurring biological entities, it would be difficult to make testable predictions or to link results obtained from systems at various levels (e.g., the genetic and the phenotypic).

Although other subfields within biology focus on variation, the modern-day genomic biosciences seek in the first instance to control such variation by imposing limits on what count as data, in hopes of understanding the fundamental mechanisms and norms and perhaps ultimately by comparison even the variations.

However, in this research what counts as a fiction shifts according to context, particularly depending on the questions being asked; so, for instance, to build digital organisms, some particular descriptions of the wiring diagrams, cell lineages, and so on must be assumed. It is understood by most of their users that these descriptions are relatively arbitrary, but they form the best available resource for doing further research, which is a form of 'useful fiction.' What started explicitly as fiction then takes on the status of a fact-like platform from which further research can be done. The ultimate goal remains the same: to project knowledge gained back to naturally occurring organisms. However, there are clear limits on the ease with which these inferences can be made, as many simplifying assumptions have been built into the systems and may become noneliminable without considerable investment in recreating new resource materials that rely on different background assumptions.

In summary, model organism research relies on replacing actual, natural entities with comparatively fictitious ones as the objects of research. Epistemic space is created by layering these fictions, and for various purposes progressively accepting various components as fact-like for the purposes of pursuing particular research goals. Thus what counts as a fictitious entity is never absolute, but a comparative, contextual matter. In order to come to a fuller understanding of various stages in scientific practice, it is important to excavate these types of fictions, as they often come to be forgotten or invisible once they become essential to the assumed data and shared practices within a field that aims to establish a true understanding of the world around us.[26]

NOTES

1. Edward Albee, quoted in Lester (1966, p. 113).
2. On literature as a source of these different types of knowledge, in addition to the works discussed in detail below, I have also found the following helpful to consider: Walsh (1969); Putnam (1978); Wilson (1983); Nussbaum (1990); Hagberg (1994).
3. John (1998).
4. Novitz (1987, p. 119).
5. John (1998, p. 335).
6. Oja (1988, p. 116); see also White (1984).
7. See Ankeny (2001a).
8. The literature in the history and philosophy of science on the development of and research with various model organisms is now voluminous, but notable contributions for the themes discussed in this essay include the essays in the collection edited by Clarke and Fujimura (1992); articles in the 1993 special issue of the *Journal of the History of Biology* on experimental organisms

edited by Burian; Kohler (1994); Bolker (1995); Schaffner (2000); Creager (2002); and Rader (2004).

9. Ankeny (2000).

10. Gibson et al. (2008).

11. Fraser et al. (1995).

12. Bono et al. (1998).

13. Glass et al. (2006).

14. Clyde A. Hutchinson, quoted in Wade (1999, p. 1).

15. I focus here on a subset of digital organisms, namely, simulation systems used to explore processes in model organisms, and not on artificial life or evolutionary biological research that uses self-replicating computer programs that mutate and evolve; for a review of this type of research, see Wilke and Adami (2002).

16. An early articulation of this approach can be found in Sejnowski, Koch, and Churchland (1988).

17. For a fairly simple example of the touch sensitivity neural circuit, see Cangelosi and Parisi (1997), which builds on the approach and data presented in Chalfie et al. (1985).

18. Kitano et al. (1988).

19. Sejnowski et al. (1988, p. 1305).

20. For more details on this process with wiring diagrams, see Ankeny (2000), on which this section draws; see also de Chadarevian (1998).

21. White et al. (1986, p. 7).

22. Ankeny (2001b).

23. Brenner (1988).

24. For a discussion more generally of the limits of model organisms in developmental biology, see Bolker (1995).

25. Eco (1983, p. 472).

26. Although I do not have space to explore this issue, this provides a clear instance where research in the philosophy of science without close attention to history and to scientific practice is likely to be misleading, as theories often rely on these sorts of hidden fictions.

12 Representation, Idealization, and Fiction in Economics

From the Assumptions Issue to the Epistemology of Modeling

Tarja Knuuttila

> I have no fellow-feeling with those economic theorists who, off the record in seminars and conferences, admit that they are only playing a game with other theorists. . . . Count me out of the game. At the back of my mind, however, there is a trace of self-doubt. Do the sort of models I try to build really help us to understand the world? (Robert Sugden, 2002 p. 107)

1. INTRODUCTION

The most persistent philosophical problem of economics has concerned the realisticness of economic theories and their basic assumptions. This so-called realism of the assumptions issue has been motivated by the seeming falsity of the basic postulates of economic theory, which has raised doubts concerning the very status of economic science. Among the contested assumptions have been those of utility maximization, perfect information, perfectly competitive markets, and the givenness of tastes, technology, and the institutional framework. The issue has been whether such assumptions are too unrealistic, or perhaps realistic enough, or whether that should matter at all.

What makes the assumptions issue an especially interesting context within which to study idealization and the use of fiction in science is the peculiarity of economics as a discipline. In terms of method, economics lies in between the natural and the social sciences. On the one hand, core economic theory has been axiomatized and economists use sophisticated mathematical methods in modeling economic phenomena. On the other hand, economics shares a hermeneutic character with other social sciences. It is largely based on everyday concepts, and as economic agents ourselves we have a reasonably good pre-understanding of economic phenomena. Our knowledge of economics feeds back into our behavior and to the functioning of economies, thus giving economics a reflexive character quite unlike the natural sciences.

Given this dual character of economics, the use of abstractions, idealizations, and fiction in economic modeling have seemed especially problematic both for economists and for the general public. The feasible economic experiments (concerning consumer behavior, for instance) are of limited value, and neither is economic theory too good in predicting, with most of the conclusions being qualitative. Moreover, any realism concerning economics seems more wedded to accurate representation than in natural sciences, because the argument for entity realism making use of experimental manipulability (Hacking, 1983) cannot easily be applied to economics (see Mäki, 1996). Thus the cognitive value of economic theory and its realistic interpretation have often been tied to the possibility of accurate representation, which in turn has added to the acuteness of the assumptions issue.

In the following I review the major standpoints concerning the realism of the assumptions issue. Interestingly, a discernible major shift has taken place in the debate: Rather than concentrating on the assumptions themselves, recent discussants have turned to the epistemology of modeling. The common denominator in the current discussion is the questioning of the very problem of misrepresentation. The proposals presented seek not to construe the "misrepresentation" that is typical of economic theories and models as the foremost methodological problem to be solved. Yet, as I will show, the representationalist tendency of appealing in one way or another to accurate depiction as *the* criterion of knowledge cannot be so easily disposed of. It is still present in proposals suggesting that, despite the admission that most of the assumptions made in economic theory are either unrealistic or downright false, at least some factors it isolates should be described correctly.

As an alternative to the characteristic representationalist problematics underlying the assumptions issue, I suggest that models should not be treated at the very outset as representations of some definite target systems. Instead, they could be conceived of as purposefully constructed independent objects, epistemic artifacts, whose cognitive value is due to their epistemic productivity. From this artifactual perspective it is tempting to compare the *functioning* of scientific models to that of literary fictions, which are also purposefully constructed cultural objects. Indeed, the common thread running through the discussion on economic models and their assumptions is the appeal to their *understandability* and *credibility*. Thus I conclude my chapter by inquiring how it can be claimed that models give us knowledge in the same way as fictions do.

2. FOLK PSYCHOLOGY AND BEYOND

The earliest understanding of the nature of the basic postulates of economics relied rather straightforwardly on *introspection*. The classical economists interpreted the basic assumptions realistically as subjectively available self-evident truths concerning human nature (see Caldwell, 1982; Mäki, 1990). Although economics as a discipline changed radically during the last half of the 19th century, the classical paradigm giving way to the so-called neoclassical view[1], the attitude of economists toward the basic assumptions of economics proved more lasting. Thus as late as in 1935 Lionel Robbins stated in his classic tract, *Essay on the Nature and Significance of the Economic Science,* that the basic postulates of the theories of value, production and dynamics were "so much the stuff of our everyday experience that they have only to be stated to be recognized as obvious" (p. 79). According to Robbins's famous dictum, "Economics is the science which studies human behavior as a relationship between given ends and scarce means which have alternative uses" (1935, p. 16). In studying human behavior economists refer to the individual's subjective valuations, which economic theory can render *understandable* but which are not observable. Although Robbins granted that empirical work could be valuable in examining the applicability of the theoretical framework to various situations and in improving the "analytical apparatus," he did not believe that such work could lead to any empirical generalizations worthy to be included among the fundamental assumptions of economics. It should also to be mentioned, however, that for Robbins the assumption of rational conduct along with the accompanying assumption of perfect foresight were not among the fundamental postulates of economic theory: the "Economic Man" was to be regarded only as an "expository device" or "first approximation" (1935, p. 96).

This attitude to economic theory and practice was soon to change, however, as the wave of positivism extended also to the methodological discussion in economics. Terence W. Hutchison (1938) provided an influential formulation for the new empiricist tenets, arguing that to have any empirical content the theoretical propositions of economics should be "conceivably" capable of empirical testing or at least reducible to such propositions. In contrast to Robbins, he considered the "fundamental principle" of economic theory to be the maximizing behavior of economic agents (consumers and firms), yet he pointed out that it relied on the unrealistic assumption that agents had full information on all economically relevant factors. When the major economic problem faced by the real agents concerns the question of how to make rational decisions in a state of uncertainty, the neoclassical solution of assuming this problem away seems a highly unsatisfactory approach. As a solution Hutchison suggested, in an empiricist fashion, that extensive empirical studies should be carried out concerning how people actually form their expectations.

That the economists so willingly embraced the positivist and empiricist message of Hutchison was also, of course, connected to the increasing mathematization and quantification of economic theory. Several important technical and conceptual advances shaping the discipline had already been set in motion both inside and outside of economics. The contributing factors included the collection of economic statistics, the development of linear programming and statistical and econometric methods, the fruitful mathematization of consumer theory, and the interventionist policy implications of the so-called Keynesian revolution and the consequent theoretical and empirical investigations of market failures and externalities. According to the general perception, economics was to become more objective and value-free, in other words a more scientific discipline (Caldwell, 1982, pp. 115–116).

3. INSTRUMENTALISM AND FICTIONALISM

The core assumptions in economics—the fundamental postulates of neoclassical microeconomics—did not change as much as how they were justified, however. The 1950s saw two new influential defenses of these basic assumptions, *instrumentalism* and *fictionalism*, both of which, in fact, denied that economic theory and its parts should be interpreted realistically. This was the underlying message in Milton Friedman's "The Methodology of Positive Economics" (1953), which became the single most read work on the methodology of economics, its very fame testifying to its success in capturing some basic convictions held by economists. According to Friedman, the "unrealism" of the assumptions does not matter. The goal of science is the development of hypotheses that give "valid and meaningful" predictions about phenomena. However, as testing a theory through its predictions is especially difficult in social sciences, this has led economists into "purely formal or tautological analysis" or, what is even worse, to evaluating the theories according to the realism of their assumptions. In doing so they are on the wrong track, because:

> The relation between the significance of the theory and the "realism" of its "assumptions" is almost the opposite. . . . Truly important and significant hypotheses will be found to have assumptions that are wildly inaccurate descriptive representations of reality and, in general, the more significant the theory, the more unrealistic the assumptions. (Friedman, 1953, p. 14)

In addition to asserting that the most important theories in various sciences are, in fact, characterized by descriptively inaccurate assumptions, Friedman also justifies his position by invoking an "as-if" formulation of economic theory. He claims, for instance, that firms can be treated "as if" they were maximizing profit in perfectly competitive markets. At this point, however, the reader starts to suspect that the essay is directed against certain new theories of competition and the firm that make more realistic assumptions than neoclassical theory concerning the structure of markets and the behavior of agents. This, in turn, rather paradoxically appears to reveal Friedman's strong belief in the basic correctness of some fundamental assumptions in neoclassical economics. Yet, at the same time, the examples through which he illustrates his "as-if" conception seem to show that he gives economic theory the status of a useful piece of fiction.

A more straightforward and consistent treatment of the role of fiction in economics is given by Fritz Machlup, who apart from attacking the ultra-empiricists (1955)—specifically Hutchison—has in his methodological writings also considered the nature and role of economic agents in economic theory.[2] Concerning *Homo oeconomicus* he observes that little consensus exists as to its "logical nature," the alternatives ranging from a priori statements and axioms to useful fictions. His own suggestion is that *Homo oeconomicus* should be regarded as an ideal type, by which he means that it is a mental construct, an "artificial device for use in economic theorizing," the name of which should rather be *Homunculus oeconomicus,* thus indicating its man-made origins (Machlup, 1978, p. 298). As an ideal type *Homo oeconomicus* is to be distinguished from real types. Thus economic theory should be understood as a heuristic device for tracing the predicted actions of imagined agents to the imagined changes they face in their environment. According to Machlup, economists are not interested in all kinds of human behavior related to business, finance, and production; they are only interested in certain reactions to specified changes in certain situations: "For this task a *homunculus oeconomicus,* that is, a postulated (constructed, ideal) universal type of human reactor to stated stimuli, is an indispensable device" (Mchlup, 1978, p. 300).

Machlup also treats neoclassical firms likewise: they should not be taken to refer to real enterprises either. According to traditional price theory, a firm is only "a theoretical link" that is "designed to explain and predict changes in observed prices . . . as effects of particular changes in conditions (wage rates, interest rates, import duties, excise taxes, technology, etc)." (Machlup, 1967, p. 9). To confuse such heuristic fiction with any real organization would be to commit "the fallacy of misplaced concreteness." The justification for modeling firms in the way neoclassical microtheory does lies in the purpose for which the theory was constructed. In explaining and predicting price behavior, only minimal assumptions concerning the behavior of the firm are needed if it is assumed to operate in an industry consisting of a large number of enterprises (in which case the effect of one firm on the aggregate output of the industry is negligible). In such a situation there is no need to talk about any internal decision making because a neoclassical firm, like a neoclassical consumer, just reacts to the constraints of the environment according to a preestablished behavioral—in other words, maximizing—principle. Yet Machlup reminds us that the motivational assumptions concerning the economic agents should be "understandable" in the sense that we could imagine reasonable people and firms acting according to them at least on some occasions.

4. REALISM REHABILITATED

As opposed to taking the seeming "unrealism" of the basic assumptions of economics at face value and then denying the importance of the realist interpretation of economic theory altogether, the realist strategy for saving the theory has been to claim that it strives to abstract real causal factors at work in real economies. One interesting proposal along these lines was presented by Uskali Mäki, who set out to deconstruct the entire problem concerning the realism of assumptions. At the heart of this problem, he claims, there is the idea that "realists prefer realistic assumptions to unrealistic assumptions, while non-realists are either indifferent or have their preferences the other way around" (Mäki, 1994, p. 239). Thus the problem concerning the realism of assumptions derives largely from the way *realism* and *realisticness* tend to be conflated (e.g., Mäki, 1990, 1992, 1994). For Mäki, "realism" denotes a family of overall philosophical tenets, whereas "realisticness" and "unrealisticness" concern the various properties of specific representations such as economic theories and their parts. Consequently, instead of assuming that in positing *unrealistic* assumptions economists adopt an *unrealist* attitude toward the economic theory, we should study the function of those assumptions.[3] In his work Mäki aims to show that unrealistic assumptions can even be the very means of striving for the truth.

The discussion concerning the realism of assumptions in economics assumes rather naively, Mäki claims, that theoretical assumptions could be too realistic. Opposing this view, he seeks to rehabilitate the method of isolation, an old and venerable idea among economists from J. S. Mill, Karl Marx, and Alfred Marshall onward. Mäki sees economic theory as an outcome of the method of isolation, in which a set of elements is theoretically removed from the influence of other elements in a given situation (Mäki, 1992, p. 318). This can be defined as follows:

> In an isolation, something, a set X of entities, is "sealed off" from the involvement of influence of everything else, a set of Y entities, together X and Y comprise the universe. The isolation of X from Y typically involves a representation of the interrelationships among the elements of X. Let us call X the isolated field and Y the excluded field. It should be obvious that any representation involves isolation. (1992, p. 321)

Thus a representation could be regarded as "unrealistic" if it "isolates a very small set of features from a very large set of features" (1992, p. 321). Mäki (1992) claims that notions in philosophical literature such as idealization, abstraction, isolation, simplification, and generalization often go undistinguished. For him, isolation is the central notion, and abstraction is a subspecies that isolates the universal from particular exemplifications. Idealizations and omissions, in turn, are techniques for generating isolations, idealizations being deliberate falsehoods, which either understate or exaggerate to the absolute extremes. It is not difficult to find examples of such idealizations in economics, where assumptions such as perfect knowledge, zero transaction costs, full employment, perfectly divisible goods, and infinitely elastic demand curves are commonly made.

As a defender of the realist interpretation of economic theory, Mäki wishes to make the point that a theory may be true even if it is partial and involves idealizations. He seeks to do away with the problem of misrepresentation by showing how a seemingly false theory may be true if it has succeeded to represent the workings of isolated causal factors in an appropriate way. Mäki (1992) puts this as boldly as stating that "an isolating theory or statement is true if it correctly represents the isolated essence of the object" (p. 344). However, in his later writings he considerably plays down his earlier claims by maintaining more moderately that truth claims should be located *inside* the models and theories concerning only the isolated forces or mechanisms (Mäki, 2004).

5. THEORETICAL VERSUS EXPERIMENTAL ISOLATION

The intuitive persuasiveness of the method of isolation is due in part to its resemblance to experimental practice. Mäki (1992) draws a parallel between "theoretical isolation" and "material isolation," referring with material isolation to experiments, which are also based on sealing off other intervening elements. This analogy becomes even more apt when instead of theories it is models that are likened to experiments—as Mäki has done in his later work (e.g., Mäki, 2005). Namely, it is possible to see modeling as well as experimentation as involving the manipulation of elements, or interventions, in controlled environments. Thus it is hardly surprising that alongside the recent interest in models the idea of models as "laboratories of economic theorists" (Mäki, 2005, p. 308; see also Hausman, 1990) has evoked considerable interest among economists and philosophers of economics (e.g., Lucas, 1980; Mäki, 1992; Cartwright, 1999b; Boumans & Morgan, 2001; Morgan, 2003, 2005). The analogy between modeling and experimentation has perhaps been most clearly spelled out by Cartwright (1999b). She compares economic modeling to what she, following Ernan McMullin (1985), calls Galilean experiment, the aim of which is to eliminate all other possible causes in order to establish the effect of one cause operating on its own (Cartwright, 1999b, p. 11). Taking an isolationist point of view we can thus approach economic models as instances of *Galilean idealization*: They purport to study how the cause operates on its own unimpeded by other causes, and to achieve this they use assumptions to neutralize the effect of other things (see also Mäki, 2005, p. 308).

However tempting and revealing the analogy between modeling and experimentation may be, the seeming easiness with which the theoretical economist as opposed to the experimentalist can supposedly effect controls by simply assuming away all disturbances and complications casts doubt on how far the analogy can go—and how feasible the method of isolation is. First, there is the problem that the "causal structure" of the real world is often such that the causes are not separable and thus may not vary independently as the method of isolation assumes (Boumans & Morgan, 2001, p. 16). This implies that we cannot have the same control of experiments as we can of our models, which is one reason why models cannot supply as powerful evidence as experiments can (Morgan, 2005, p. 323).

Secondly, even though the theorist appears to be more free to effect controls than the experimentalist, she faces in her modeling task different kinds of constraints, due to the representational means used. For instance, mathematical modeling, which is by far the most popular tool of modern economists, is driven by the requirements of tractability. These requirements are especially stringent in economic modeling partly because economists typically seek analytically solvable equations (see Lehtinen & Kuorikoski, 2007). Thus, although Mäki claims that "a realist has to employ unrealistic assumptions to get to the truths about limited but causally significant aspects of reality" (Mäki, 2005, p. 1731), Cartwright is more cautious in terms of how this is supposed to succeed in economic modeling. Although she was an earlier defender of the isolationist strategy in establishing causal tendencies or capacities (e.g., Cartwright, 1998), she has recently questioned whether economic models are "overconstrained" to enable the study of the isolated processes of interest.

The specific problem with economic models is that many of their idealizations are not meant to shield the operation of the causal factor or tendency of interest from the effects of other disturbing forces. Rather, they are made for reasons of tractability: The model economy has to be attributed very special characteristics so as to allow such mathematical representation that, given some minimal economic principles such as utility maximization, one can derive deductive consequences from it (Cartwright, 1999b, p. 7). However, this tends to water down the idea that as the investigation proceeds one could achieve more realistic models through de-idealization: It is difficult to make sense of the very idea of relaxing assumptions that are aimed at facilitating the derivation of the results from the model (Alexandrova, 2006). One simply does not know how statements concerning such "derivation facilitators" should be translated into statements about the real entities and properties:

> In what sense is it more realistic for agents to have discretely as opposed to continuously distributed valuations? It is controversial enough to say that people form their beliefs about the value of the painting or the profit potential of an oil well by drawing a variable from a probability distribution. So the further question about whether this distribution is continuous or not is not a question that seems to

make sense when asked about human bidders and their beliefs. (Alexandrova, 2006, p. 183)

Thus the assumptions needed to make the model mathematically tractable threaten the very idea of isolation, because then the problem concerns not only the unrealisticness of the assumptions but also the model dependence of the results derived. This, claims Cartwright, is a typical problem of economic models, which are characteristically based on "thin" everyday concepts and few general theoretical principles, being, for this very reason, full of other structures. Cartwright cites as an example the characterization of a model constructed by Robert Lucas, Jr., to study the "money illusion" (Lucas, 1981). Lucas starts his discussion by describing an "abstract model economy" in which in each period N identical individuals are born, each of them living for two periods and in which in each period there is then a population of $2N$: N of age 0 and N of age 1. He then goes on to specify the output and consumption of that economy as well as the manner in which money is created and the exchange takes place in it. His description of the model economy takes several pages and the assumptions made are patently artificial in the sense that it is difficult to imagine how they could have been drawn from the economic reality (even with the help of suitable idealizations). They are clearly purpose-built assumptions that provide a way to secure "deductively validated" results, yet at the same time they tie the results to the model-specific circumstances created (Cartwright, 1999b, p. 18).[4]

Cartwright's skepticism concerning the "vanity of rigor in economics" is partly backed up by the comparison of economics to physics, in which in many cases comprehensive and well-confirmed background theories exist, giving the resources with which to estimate the effect of distortions introduced by specific idealizations, and providing guidance on how to attain particular levels of accuracy and precision. However, to do justice to economics it should rather be compared with sciences such as ecology and population biology that strive to model complex phenomena lacking any comprehensive foundation.[5] In such cases the models can be at least partly justified by what several analysts following Richard Levins (1966) have called "robustness analysis" (Wimsatt, 1987; Weisberg, 2006). Robustness can be characterized as stability in a result that has been determined by various independent means, for instance through observation, experiment, and mathematical derivation. Applied to modeling, robustness means the search for predictions common to several independent models (Weisberg, 2006). However, in economics something like robustness analysis is carried out on models that are far from independent, usually being variations of a common "ancestor" and differing from each other only with respect to a couple of assumptions. Thus it is possible to claim that by constructing many slightly different models economists are in fact testing whether it is the common core mechanism of the group of models in question that is responsible for the result derived and not some auxiliary assumptions used (Kuorikoski et al., 2007). Robustness analysis is a way of finding out which assumptions are the ones that are "doing the work" (Sugden, 2002, p. 129).

Consequently, robustness analysis seems to be at least a partial answer to the question of whether or not the results derived from economic models are due to their assumed core causal mechanism. Yet this form of reasoning remains within the realm of modeling, and making it empirically relevant seems to hang ultimately on whether or not the group of models in question has really succeeded in isolating and adequately representing a real mechanism at work. A closer look at many economic models, such as Lucas's money illusion model, reveals that it is hard to conceive of them as products of isolation effected by suitable idealizations. Thus robustness analysis alone, at least in the form practiced by economists, does not manage to dispel Cartwright's qualms concerning the isolationist strategy in economics.

It seems to me that fundamental questions concerning the nature of modeling and the way they are supposed to give us knowledge are at stake here. If anything, models are considered by philosophers of science to be representations (e.g., Hughes, 1997; Suárez, 1999; Teller, 2001; Frigg, 2003), and the main difference between them and experiments concerns the fact that they are constructed by representing: When "experimenting" with models one is manipulating representations rather than any real entities and processes the experiment is supposed to be about, as in laboratory experiments. The isolationist account sidesteps this problem by simply assuming that one can represent at least the workings of one causal factor or tendency correctly, and then as a result of suitable idealizations study its behavior. Yet, as indicated, this does not seem to effectively depict what happens in economic modeling. First, even if it were possible to neatly identify and separate the real causal factors of interest in the way assumed by the method of isolation, there is the problem that the model assumptions do not just neutralize the effect of the other causal factors. They do much more: They construct the modeled situation in such a way that that it can be conveniently mathematically modeled. Moreover, in this process such properties are attributed to the modeled entities and their behavior that the model starts to look like an intricate construction rather than a neat model experiment involving isolations and idealizations according to some well-regimented scheme. How else, then, could one approach economic models?

6. CREDIBLE PARALLEL REALITIES

As I have already indicated, the crucial question concerning the method of isolation is: Do economists really go on isolating causally relevant factors of the real world and then use deductive reasoning to work out what effects these factors have in particular controlled (model) environments? According to Robert Sugden, who is a theoretical economist himself, this does not match the theoretical practice of economists. In Sugden (2002) he investigates how model building can tell us something about the real world by conducting a close analysis of two influential economic models, while also taking heed of what their authors claim. The cases studied are George A. Akerlof's 1970 "market for 'lemons,'" which is famous for introducing the concept of asymmetric information, and Thomas Schelling's "checkerboard model" of racial sorting taken from his book *Micromotives and Macrobehaviour* (1978). In the light of these cases, Sugden reviews several standpoints taken in the methodology of economics, although he puts most of his effort into discussing the method of isolation as presented by Mäki and "the economists' inexact deductive method" developed by Daniel Hausman (1992). These accounts provide him with the main target of his criticism. He finds them similar in that both consider the assumptions in economics to be usually very unrealistic, yet the operations of the isolated factors should be described correctly.

As opposed to this, Sugden claims that economic models should rather be regarded as constructions, which instead of being abstractions from reality are *parallel realities*. A good example toward that end is the "checkerboard model," which Schelling uses to explain segregation by color and by sex in various social settings. Schelling suggests that it is unlikely that most Americans would like to live in strongly segregated areas, and that this pattern could be established only because they do not want to live in a district in which the overwhelming majority is of the other color. The model consists of an 8×8 grid of squares populated by dimes and pennies, with some squares left empty. In the next step a condition is postulated that determines whether a coin is content with its neighborhood: It might, for instance, be content if one third of the coins adjacent to it are of the same type. Whenever we find a coin that is not content we move it to the nearest empty square, despite the fact that the move might make other coins discontented. This continues until all the coins are content. As a result, strongly segregated distributions of dimes and pennies tend to appear—even if the conditions for contentedness were quite weak.

According to Sugden, it seems rather dubious to assume that a model like the checkerboard model is built by representing some key features of the real world and sealing them off from the potential influence of other factors at work: "Just what do we have to seal off to make a real city—say Norwich—become a checkerboard?" he asks (Sugden, 2002, p. 127). Thus, "the model world is not constructed by starting from the real world and stripping out complicating factors: although the model world is *simpler* than the real world, the one is not a simplification of the other" (p. 131). Rather than considering models as isolating representations he prefers to treat them as constructions, with the checkerboard plan being something that "Schelling has *constructed* for himself" (p. 128). What are these models then constructed for, and more specifically, how are they used to gain knowledge about the real world? What puzzles Sugden is that even though both Akerlof and Schelling are willing to make empirical claims about the real world based on their models, it is difficult to find from their texts any explicit connections made between the models and the real world (p. 124). Their claims about the real world are not the ones they derive from their models, but something more general. Thus there is a seeming gap in their argumentation that makes Sugden ask how the transition from a *particular* hypothesis, which has been shown to be true in the model, to a *general* hypothesis concerning the real world can be justified.

In answering this question, Sugden, a model builder himself, suggests that modelers are making *inductive inferences* on the basis of their models. One commonly infers inductively from one part of the world to another, expecting that the housing markets of Cleveland resemble those of other large industrial cities in the northeastern United States, for instance. However, just as we can infer from real systems to other real systems, we can also infer from theoretical models to "natural models." A modeler constructs "imaginary cities, whose workings he can easily understand" in order to invite inferences concerning the causal processes that might apply to real multiethnic cities. This possibility is based on our being able to see the relevant models as instances of some category the other instances of which might actually exist in the real world (p. 130). For this to work we have to accept that the model describes a state of affairs that is *credible* given our knowledge of the real world, and in doing so it could be considered realistic in much the same fashion as a novel can. Even though the characters and the places in the novel might be imaginary, we could consider them credible in the sense that we take it to be possible that there are events that are outcomes of people behaving as they do in the novel.

7. THE TURN TO MODELING

Sugden's article on credible worlds provides one of the clearest examples of the change that has happened with regard to the question of the (un)realism of the fundamental assumptions in economics. The discussants no longer worry about the unrealisticness of the assumptions per se, as the methodological interest has turned toward the construction and functioning of economic models.[6] What seems especially interesting in Sugden's account is his suggestion that the (supposed) causal factors are built into the specification of a model, and yet he denies that a model is constructed primarily by isolating some causal factors that are operative in the world and then abstracting away from the interference of other factors. Yet, in spite of his claims, it might be tempting to place him in the isolationist camp (cf. Alexandrova, 2006). Although I do not find this interpretation of Sugden appropriate, it is relatively easy to pinpoint the reason for it: namely, the proponents of the isolation account hardly contest the fact that models are constructed (see Mäki, 2005). Moreover, because they posit that what is being isolated are some causal factors (Mäki, 1992, 2004) or causal tendencies or capacities (e.g., Cartwright, 1998), it may be difficult to see how Sugden's account, *pace* his claims, differs from theirs.

In order to see where exactly Sugden parts company with the isolationists I suggest taking a closer look at *what* is actually being represented in economic models. It is somewhat paradoxical that most philosophical discussion on scientific representation so far has been conducted in the context of modeling, whereas scientific endeavor employs manifold representations that are not readily called models. Such representations include visual and graphic displays on paper and on screen such as pictures, photographs, audiographic and three-dimensional (3D) images, as well as chart recordings, numerical representations, tables, textual accounts, and symbolical renderings of scientific entities such as chemical formulas. In fact, modeling presupposes these more "basic" forms of representation, which organize the data and present theoretical objects and fields of interest in some representational medium. Moreover, especially in the case of simulation, models provide the medium for producing further representations. This makes one ask whether modeling as a distinct scientific practice perhaps nurtures a very specific kind of representation. The preceding discussion on economic models and especially on Sugden's claims questions the assumption that such models could be understood as being derived by representing some features of real target systems. Yet, as models they are certainly representations in the sense that they are able to convey content through the use of conventional representational means such as mathematical formalisms. How should this special representational nature be understood?

In an attempt to define the nature of modeling, Michael Weisberg (2007) suggests that it could be distinguished from other forms of theorizing through the procedures of *indirect* representation and analysis that modelers use to study real-world phenomena (see also Godfrey-Smith, 2006, for a similar view). Weisberg contrasts Vito Volterra's style of theorizing, which he takes as an example of modeling, to the "abstract direct representation" as exhibited by Dimitri Mendeleev. Volterra studied the special characteristics of post-World War I fish populations in the Adriatic Sea by "imagining a simple biological system composed of one population of predators and one population of prey" to which he attributed only a few properties, writing down a couple of differential equations to describe their mutual dynamics. The word "imagining" used by Weisberg here is important because it captures the difference between the procedures of direct and indirect representation. He stresses the fact that Volterra did not arrive at these model populations by abstracting away properties of real fish, but rather constructed them by stipulating certain of their properties (p. 210). Unlike Volterra, he claims, Mendeleev did not build his Periodic Table via the procedure of constructing a model and then analyzing it. In developing his classification system—which according to Weisberg should also be seen as a theoretical achievement—he was rather working with abstractions from data in an attempt to identify the key factors accounting for chemical behavior. Thus, in contrast to modelers such as Volterra, Mendeleev was trying to "represent trends in real chemical reactivity, and not trends in a model system" (p. 215, footnote 4).

Coming back to economic models, I would suggest that making a distinction between abstract direct representation and indirect representation is fruitful in locating where the difference between isolationists such as Mäki and (earlier) Cartwright and constructivists such as Sugden and Lucas lies. The latter are considering economic theory in terms of modeling. The crucial difference between isolationists and modelers is not about whether one abstracts or approximates, selects or even idealizes. Scientific representation involves all these, but in engaging in such activities, modelers, the true constructivists, do not even pretend to be primarily in the business of representing any real target system. For them the models come first. The distinguishing feature of the "strategy of model-based science" is that the modelers do not attempt to identify and describe the actual systems, but proceed by describing another more simple hypothetical system (Godfrey-Smith, 2006). Thus model-based science could be characterized by the "deliberate detour through merely hypothetical systems" of which it makes use (Godfrey-Smith, 2006, p. 734).

What is typical of modeling is that models provide the modelers with independent objects that are often investigated in their own right (Godfrey-Smith, 2006; Knuuttila, 2005; Weisberg, 2007). They might be interpreted as models of some external systems and assessed according to their relationship with the world, but they need not be: As independent objects also studied for their own sake they need not have any predetermined representational relationship with the real world (Suárez, 2004b; Knuuttila & Voutilainen, 2003). Different modelers may use the same model in different ways, and thus also assess its potential representational relationships with real target systems differently (Knuuttila, 2005; Godfrey-Smith, 2006; Weisberg, 2007). What is more, modelers are often interested in clearly nonexisting phenomena such as three-sex biology—and this also applies to economic models, which are often blatantly unrealistic and too contrived to qualify for anything that could seriously be taken to be exhibited in real life. In these cases modelers clearly seem to be trading with fiction, to which point I return later. Last but not least, the fact that models could be considered independent objects of study in themselves explains why their supposed relationships with real systems frequently remain implicit or loose, with the modelers often being most interested in understanding the behavior of their model systems. This suits the practice of economists, who have repeatedly been accused of playing with their model economies instead of tackling the problems of real economies. This would not be such a pressing issue if economic models were treated first and foremost as representations of some predefined target systems.

8. SURROGATE REASONING AND THE
RIDDLE OF REPRESENTATION

The idea of models as intentionally constructed surrogate objects that are used indirectly in order to gain knowledge is by no means new (to name some recent writers on the topic, see Swoyer, 1991; Suárez, 2004b; Knuuttila & Voutilainen, 2003; Mäki, 2005). Yet Weisberg and Godfrey-Smith succeed in giving it an interesting twist by distinguishing between direct and indirect representation. At first glance, however, the very idea of making such a distinction might seem a contradiction in terms—is not reasoning from any representation always indirect, in that it makes use of a sign vehicle that is assumed to stand for the supposed target one is reasoning about? In this sense the knowledge we acquire from representation is always indirect.[7]

Indeed, this is how Mäki in his later writings approaches the question of how we reason from models. According to him, models are representations "that are *directly* examined in order to *indirectly* acquire information about their target systems" (Mäki, 2005, p. 303, emphasis added by the author). He divides the representational relationship between models and their targets into two parts, the *representative* and *resemblance* aspects. On the one hand, models are "representatives" in that they "serve as 'substitute systems' of the target systems that they represent" (Mäki, 2005, p. 304). On the other hand, being representatives they "prompt the issue of resemblance" between themselves and their target systems. Concerning resemblance, Mäki claims rather straightforwardly that "in order for a model to serve as a good representative, it must resemble its target system in relevant respects and sufficient degrees" (Mäki, 2006, p. 9).[8]

Yet the philosophical gist of the idea of indirect representation is exactly that we need not take models as representatives of any preexisting target systems *at the very outset*. By way of contrast, according to the notion of indirect representation a model can be constructed by representing *something*—and being a representation in this sense—without having necessarily to be a representation of a certain real target system or mechanism. We are perhaps not able to trace the features of a model back to any specific target system through any chain of representational transformations, idealizations, or abstractions. Instead, models are hypothetical systems typically constructed on the basis of some common and rather general "stylized features" and theoretical insights concerning the phenomenon of interest with the help of available mathematical and other representational tools.[9]

Still, in the light of philosophical tradition, this may sound counterintuitive. How can models license surrogate reasoning if they are not derived from the very real-life systems they are supposed to be about? Surely they give us knowledge exactly in their capacity as *models of* some target systems? Here, it seems to me, we arrive at the very heart of the puzzle concerning models and representation: namely, in the customary double move of treating models as representations and then ascribing their cognitive value to representation, the weight appears to be shifted to giving an adequate account of representation (see Teller, 2001). Indeed, this is what several philosophers have taken as their task in the recent discussion on models (e.g., French, 2003; Giere, 2004; Suárez, 2004b; Contessa, 2007; Mäki, in press; Frigg, in press). Yet the outcome of that discussion has been that representation has not met the bill: The pragmatist account, the most plausible analysis of representation presented so far, is minimalist to the extent of not explaining *in virtue of what* models can give us scientific knowledge and understanding.

The reason for this conclusion is clear. For the representational relationship to explain in virtue of what we can obtain knowledge through surrogate reasoning, it has to establish some sort of privileged relationship between the model and its target. By privileged relationship I mean a relationship that is somehow grounded on the very nature of both the representative vehicle and its target system, thus making them share some corresponding features or some structure. This would explain how the information gained through inspecting the surrogate could be transferred back into knowledge concerning the target. This idea is taken up in the so-called isomorphism and similarity accounts of representation, according to which the structure specified by a model represents its target system if it is either structurally isomorphic or somehow similar to it (e.g., van Fraassen, 1980; French & Ladyman, 1999; Giere, 1988).

The pragmatist critics of the aforementioned structuralist (or semantic) conceptions have argued, rather conclusively I think, that the dyadic relation of correspondence between the representative vehicle (a model) and its target assumed by the isomorphism and similarity accounts does not satisfy either formal or other criteria we might want to impose on representation (Suárez, 2003; Frigg, 2003). The dyadic conceptions attempt, as Suárez aptly put it, "to reduce the essentially intentional judgments of representation-users to facts about the source and target objects or systems and their properties" (2004b, p. 768). As opposed to this attempt, the pragmatist approaches focus on the intentional activity of representation users, denying that representation may be based on the certain properties of the representative vehicle and its target object. Yet, as a result, nothing really substantial can be said about representation in general. This has also been admitted by proponents of the pragmatic approach (cf. Giere, 2004; Suárez, 2004b), among whom Suárez has most explicitly argued for a "deflationary" account of representation that seeks not to rely on any specific features that might relate the representative vehicle to its target.

9. MODELS AS EPISTEMIC ARTIFACTS

With the pragmatist notions of representation we have thus reached the limits of the representationalist paradigm in terms of explaining how models can give us knowledge. Representationalist thinking is grounded on the idea that knowledge consists of representations that correspond accurately, or in practice at least partially, to the reality as it is, quite apart from our cognitive activities and abilities or the available representational tools. Nevertheless, as argued earlier, this ideal does not stand closer scrutiny, if only because representation has not been given, to date, any satisfactory substantial analysis. The pragmatist account, in turn, paradoxically gives us an antirepresentationalist account of representation in seeking no deeper constituent features that might relate the representative vehicle (the source) to its target. It is nevertheless important to be clear on what is established by the pragmatist account. In fact, it just points to the impossibility of giving a general *substantial* analysis of representation that would explain in virtue of what models give us knowledge concerning real target systems. It does not deny that many scientific representations can be traced back to some target objects, or that they can depict them more or less accurately at least in some respects, the clearest cases of such models being scale models and maps. Yet, as we saw earlier in the case of economic models, it is highly contestable whether models in general can be treated as partial representations (i.e., isolations) of some real target systems or mechanisms.

In this situation, rather than trying to give yet another overall philosophical analysis of representation—I do not feel uncomfortable settling for the minimalist account; quite the contrary, it gives me the relief of finally opening the black box and finding it empty—I suggest that we also look for other ways of approaching models. The preceding discussion has given two clues for pursuing such a strategy: first, the idea that models are independent objects, and second, the analogy of models to fictions. With enough hindsight it is possible to see how these paths eventually lead in the same direction and may partly converge, but to start from that would entail a rapid loss of some important insights concerning the epistemic value of modeling. Thus, I first consider what insights are inherent in the idea of models as independent objects.

What I find the most important point in viewing models as independent entities is that it enables us to appreciate their productive characteristics. Adopting a productive perspective requires one to address them as autonomous but also *concrete objects* that are constructed for *epistemic purposes* and whose cognitive value derives largely from our *interaction* with them. This was recognized by Morrison and Morgan (1999; Morrison, 1999), who focus in their account of models as mediators on how we learn from them by constructing and manipulating them. This, they claim, is made possible by the model's partial autonomy, which they interpret in terms of its relative independence from the traditional theory–data framework according to which "models are *not* situated in the middle of an hierarchical structure between theory and the world" (p. 17) However, it seems to me that their view permits, and to be efficient in fact requires, a more radical reading. If our aim is to stress how models enable us to learn from the processes of constructing and manipulating them, it is not sufficient that they are considered autonomous: They also need to be concrete in the sense that they must have a tangible dimension that can be worked on. Such concreteness is provided by their material embodiment, which gives them the spatial and temporal cohesion that enables their maneuverability. This also applies to so-called abstract models: When we are working with them we typically construct and manipulate external diagrams and symbols. Thus, even abstract entities need to have a material dimension if they are to be able to mediate. Herein also lies the rationale for comparing models to experiments: In devising models we construct self-contained artificial systems through which we can articulate our theoretical conjectures and make them workable and "experimentable" (see Rouse, this volume).

In terms of productivity, any privileged "underlying structures" alone—which form the core of the semantic conception of models—do not take us very far. To wit, we only need to consider the different kinds of models used: scale models, diagrams, different symbolic formulas, model organisms, computer models, and so on. This very variation suggests that their material dimension and the diverse representational media they make use of are crucial for promoting understanding through them. Consequently, it could be claimed that one should take into account the medium through which scientific models are materialized as concrete, intersubjectively available objects (Knuuttila & Voutilainen, 2003; Knuuttila, 2005). This media-specific approach focuses on their constraints and affordances, which are partly attributable to the characteristics of the specific representational means (pictorial, diagrammatic, symbolic, 3D, etc.) with which they are imbued. From this perspective the use of representational media provides external scaffolding for our thinking, which also partly explains the heuristic value of modeling.[10] It is already a cognitive achievement to be able to express any mechanism, structure, or phenomenon of interest in terms of some representational media, including assumptions concerning them that are often translated in a conventional mathematical form. Such articulation enables further arguments as well as theoretical findings, but it also imposes its own demands on how a model can be achieved—as illustrated earlier regarding the intricacy of mathematical modeling in economics. In this sense any representational media is double-faced in both enabling and limiting.

Another aspect of scaffolding provided by models is related to the way in which they enable us to draw many kinds of inferences from them as they help us to think more clearly and to proceed in a more systematic manner (see Suárez, this volume). Models are typically constructed in such a way that they constrain the problem at hand, thereby rendering the situation more intelligible and workable. As the real world is just too complex to study as such, models simplify or modify the problems scientists deal with. Thus, modelers typically proceed by turning the constraints (e.g., the specific model assumptions) built into the model into affordances; one devises the model in such a way that one can gain understanding and draw inferences from using or "manipulating" it. This experimentable dimension of models accounts for how they can have explanatory power providing a framework for various what-if-things-had-been-different questions (Bokulich, this volume).[11] (Yet their seeming simplicity often disguises their very heterogeneity in terms of the various elements they are made of, such as familiar mathematical functions, already established theoretical entities, certain generally accepted solution concepts, and so forth.) Hence the very construction of models explains, on the one hand, how they allow *certain* kinds of solutions and inferences, and on the hand how they can also lead to unexpected findings, thereby breeding new problems and opening up novel areas of research.

I suggest that we gain knowledge through models typically by way of building them, experimenting with them, and trying out their different alternative uses—which in turn explains why they are regularly valued for their *performance* and their *results* or *output*. From the productive perspective, rather than trying to represent some selected aspects of a given target system modelers often proceed in a roundabout way, seeking to build hypothetical model systems in the light of the anticipated results or of certain general features of phenomena they are supposed to exhibit. If a model succeeds in producing the expected results or in replicating some features of the phenomenon, it provides an interesting starting point for further theoretical conjectures and inferences. This result-oriented characteristic also accounts for why modelers frequently use the same cross-disciplinary *computational templates* (Humphreys, 2002), such as well-known general equation types, statistical distributions, and computational methods. The overall usability of computational templates is based on their generality and the observed similarities between different phenomena. Thus there is an element of opportunism in modeling: The template that has proven successful in producing certain features of some phenomenon will be applied to other phenomena, often studied by a totally different discipline. The aim of getting the model to bring forth results also explains why tractability considerations frequently override the search for correct representation (Humphreys, 2004).

Consequently, the very peculiarity of the epistemic strategy of model-based science lies in the fact that models are artificial entities constructed with their manipulability, experimentability, and the questions they are supposed to give answers to in mind. That this strategy should prove fruitful lies partly in the fact that we know what we have built into our models. Thus even "false," or rather fictional, models can give us knowledge, something that is indeed commonplace in scientific practice, which shows that models may also be good *inferential devices* when they do not represent real objects accurately in any relevant respect (see Wimsatt, 1987; Knuuttila, 2006; Suárez, this volume). This is typical of economic models, which present tentative mechanism sketches and function as demonstrations or exemplifications of different possibilities (see Elgin, this volume). Moreover, the productive approach allows different kinds of epistemic strategies, depending crucially on what kinds of artefacts we have at our disposal. For instance, simulations seem problematical from the representational perspective because "instead of creating a comprehensive, though highly idealised, model world, [they] squeeze out the consequences in an often unintelligible and opaque way" (Lenhard, 2006, p. 612). Yet from the productive perspective they seem to create a new kind of pragmatic understanding oriented toward control, design rules and predictions (Lenhard, 2006, p. 612).

It is worth noting that several prominent economists have expressed the kinds of views that appear to be in accord with the artifactual account I have suggested. Perhaps the most well known of these methodological statements was given by Lucas in his "Methods and Problems in Business Cycle Theory" (1980). In the famous opening of this article he claims:

> One of the functions of theoretical economics is to provide fully articulated, artificial economic systems that can serve as laboratories in which policies that would be prohibitively expensive to experiment with in actual economies can be tested out at much lower cost. To serve this function well, it is essential that the artificial "model" economy be distinguished as sharply as possible in discussion from actual economies. . . . This is the sense in which insistence on the "realism" of an economic model subverts its potential usefulness in thinking about reality. Any model that is well enough articulated to give clear answers to the questions we put to it will necessarily be artificial, abstract, patently "unreal." (Lucas, 1980, p. 696)[12]

As for the assumptions needed to build a model, Lucas states:

> A "theory" is not a collection of assertions about the behaviour of the actual economy but rather an explicit set of instructions for building a parallel or analogue system—a mechanical, imitation economy. (Lucas, 1980, p. 697)

Of course, the occasional methodological outbursts of economists such as Lucas (not to mention Friedman's classic statement) are never that innocent, and are often aimed at sponsoring a certain kind of theory or model.[13] Yet the arguments used are such that are thought to persuade fellow economists. Another prominent economist, Robert Solow (1997), made a related although slightly different point in favor of the productive view of models. Reflecting on the current state of economics, he attributed the modeling activity of economists to the increasing amount of empirical data. In his opinion, models are often devised in order to account tentatively for different empirical findings, an observation that underlines the importance of the results derived from them.

10. SCIENTIFIC FICTIONS?

The artifactual approach presented earlier in this chapter suggests that models may in many respects *function* like fictions, which are also seemingly autonomous entities constructing their own "worlds." Both Sugden (2002) and Peter Godfrey-Smith (2006), who have recently taken up fictions and imagined objects in their discussion on modeling, point out the apparent ease with which we are able to move between imagined and real-world systems. Why this should be the case they do not explicate much further. To find more about what is at stake, let us take a short trip to literary theory. Interestingly, the recent dialogue on fiction in literary science has been largely inspired by the philosophical discussion on possible worlds. However, rather than having applied the insights gained from this discussion more or less directly to the analysis of fictional worlds, literary scholars have adapted that philosophical framework to better suit their needs (e.g., Eco, 1979; Pavel, 1986; Ronen, 1994; Dolezel, 1998). It is possible to extract and extrapolate from this discussion a host of properties of fiction that also seem directly relevant when it comes to modeling. These properties could be discussed under the headings of *constructedness, autonomy, incompleteness,* and *access.*

Constructedness gives the minimal criterion for what may be regarded as fictional: Fictional worlds are constructed, and do not exist apart from having once been represented. Thus fiction contrasts at the outset with reality, which we take to exist quite apart from our representational endeavors. Fiction deals with the possible and the imaginary, with nonactual states in general, which is the reason why the fictional mode is not limited to the literary realm but extends to other discourses as well. Models are like literary fictions in that they are human-made objects, and if we are to take the proposal of indirect representation seriously, they could be considered fictional in exactly the sense that the existence of the entities, relations, and processes set up by the model worlds are due to their having been described and stipulated. Constructedness also directs our attention to the culturally available representational tools with which fictions are attained. I take it as no surprise that the question of fiction has arisen precisely in the context of mathematical modeling, because the use of mathematics and formal languages in modeling resembles the use of ordinary language in literature in that it makes use of arbitrary signs in conveying content. It would feel less natural to speak about fiction in relation to scale models or diagrams, because they make use of the iconic properties of the representational media in question.

An important characteristic of fictions that has also attracted due attention in the recent discussion on modeling is their *autonomous* or independent status. A fictional world posits a self-sufficient system of entities and their relationships, which justifies the use of the word "world"[14]: "a world is an autonomous domain in the sense that it can be distinguished from other domains identified with other sets of entities and relations" (Ronen, 1994, p. 8).[15] This relates to what I have argued earlier about models: What distinguishes models from other scientific representations is exactly their independent systemic character, which I take as crucial for their epistemic functioning. This accords with Michael Lynch's meticulous ethnographic study on scientific representations, which shows that whereas more elementary representations such as different visual displays often further fragment the object or specimen to reveal its details, models differ from them in reconstructing a "holistic entity" (1990, p. 167). Moreover, the independence of fictional worlds explains why it does not seem right to characterize fictions as falsehoods (or truths, for that matter), although one may very well make true or false claims inside the fictional world. The autonomous characteristic of fictions means that they do not make any direct claims about our actual world. It is indeed typical of them, as well as of models, that they are often studied for their own sake.

Although self-contained, fictional worlds are also *incomplete* as compared to the real world. They are "small worlds" that only contain selected features (Eco, 1990), and consequently "only some conceivable statements about some fictional entities are decidable, while some are not" (Dolezel, 1998, p. 22). The question of whether Emma Bovary had a birthmark on her left shoulder, for instance, is undecidable. To deny "incompleteness to fictional entities is tantamount to treating them as real entities" (Dolezel, 1998, p. 23). The incompleteness of fiction has consequences in terms of how it is supposed to be interpreted. Although Emma Bovary's birthmark is not the sort of thing we need to know about when reading *Madame Bovary*, in making sense of the text we nevertheless use our "world knowledge" and experience to understand her story. Our inferential activity is needed to relate the fictional world to the external cultural and factual framework— and in this sense fictional worlds are always secondary to the actual world that provides the background knowledge needed to understand the fiction (Eco, 1979; Pavel, 1986). This explains the ease with which we can travel back and forth between the fictional and the actual world: The fictional world would not be understandable in the first place unless we were able to relate it to our experience and knowledge concerning the actual world. The understanding gained through models works in a similar fashion, although we obviously need much more specific expert knowledge to interpret scientific models than to understand fiction, although different literary genres also make different demands on their readers.

Finally, as the fictional world is constructed by way of representing it, the only *access* to it is through the description of it. This, of course, does not mean that fictions any more than models should be identified with their descriptions. The story of Anna Karenina cannot be identified with a collection of letters on paper. Although once written her story became part of our cultural heritage and a possible topic in itself, in offering different interpretations of it literary critics typically refer back to the empirical text in justifying their analysis. Neither can a scientific model be reduced to an abstract entity or to any description of it. In a sense, it includes both the abstract dimension and the materially accomplished description, yet for its existence it still needs users to interpret the description according to the relevant conventions. This also accounts for the double movement that is typical of modeling and through which modelers oscillate continuously "between thinking of a model system in very concrete terms, and moving to a description of purely mathematical structure" (Godfrey-Smith, 2006, p. 736).

In drawing the preceding parallels between models and (literary) fiction, I do not wish to claim that models *are* fictions. In fact, many literary scholars think that what we consider fiction is a pragmatic matter. The point is, however, that we make sense of fictions in the same way as we make sense of factual accounts. We just make different demands of them. This explains also the "face value practice" of scientists concerning models: They are described and interpreted in the same way as real physical systems are, although they do not describe actual systems, and scientists are, for the most part at least, aware of this (Thomson-Jones, in press). Thus, economists interpret the theoretical entities postulated by economic models (consumers, firms, investments and markets, for example), making use of the concepts that are familiar to them from real-life economic contexts. This does not mean blurring the borders between fictional and factual accounts—or between literary fictions and scientific models. It is just a fact concerning how we make sense of signs and concepts in general. Models have to be interpreted if they are to make sense, and this interpretation relies on an interconnected web of different concepts—both scientific and ordinary— that is based on our world knowledge and empirical experience.

The different demands we make on literary fictions and scientific models, for instance, imply that there are important pragmatic differences between them. As I have claimed earlier, the key to the specific epistemic value of models is their purposefully constrained construction and their productive nature. To be sure, literary fictions are also purposefully constructed, but in contrast to them models are built with their "expediency in inference" (Suárez, this volume) in mind. These inferences are supposed to concern the real world, which is not what is required from literary fictions. Even if literary fictions may add to our knowledge of the actual world, as they often do, that cannot be claimed to be *the* central function of literature. Yet models need to be answerable to empirical observations and experimental findings, in other words, to other scientific practices generating knowledge about the real world. The expediency of inference is also furthered in modeling by the media it makes use of. As opposed to using natural language, modelers strive to make use of formal languages and mathematics for their very exactness and for the possibility of deriving results from them. Whereas, like other theorists and experimentalists, they strive to create what Ludwik Fleck has called "a maximum thought constraint" (1979, p. 95), the novelists and artists play with ambiguity and multiple interpretations. This does not mean that modeling cannot lead to new interpretations, however: Models are constructed in such a way as to allow further theoretical conjectures and inferences, and they are frequently reinterpreted, especially when the model templates are transferred to other areas of study.

Regardless of the differences between scientific models and literary fictions, I hope that the preceding discussion concerning the similarities of their respective functioning has nevertheless established the point that the links forged between models and some external target systems are looser than the representationalist picture would have us believe. As illustrated in the case of the models Sugden studied, the conclusions drawn from them were more general than those derived from the models themselves, and the modelers did not specify exactly what bearing their model results were supposed to have on the conclusions. There thus seemed to be a gap between the models and the claims based on them. The discussion on the functioning of fiction suggests, however, that the gap is closed by taking an inferential leap based on the relevant background knowledge. Scientific reasoning consists of a subtle triangulation of different kinds of results and evidence (from models, experiments, measurements and observations) in the light of already certified background knowledge. From this perspective the representationalist gaze, fixed on the model–target dyad and its respective properties, has assumed a unit of analysis that is too limited, something that has been criticized in pragmatist analyses of representation.

Last but not least, it appears to me that the representationalist approach accepts the cognitive challenge of modeling in the wrong way. It is as if we knew already what to isolate and had the suitable representational means at hand for doing this, thus abstracting away from the very process and means of representation. From the productive perspective, we rather engage in modeling in order to know more about the mechanisms underlying the phenomena of interest. In the face of the unknown, we try to make use of what is already known and understood, what provides the rationale for using as-if reasoning, analogies, familiar computational templates, and other constructive techniques, all of which play a central role in model-building (Fine, 1993).

11. CONCLUSIONS

I have reviewed some prominent proposals for resolving the so-called assumptions issue in economics. The issue concerns the problematic nature of idealizations and fictions in economic theories and models. As such, it does not differ from the corresponding problems in other sciences, yet the problem is heightened in economics because of its special nature as a social science with hermeneutic underpinnings that has striven to mathematize its theoretical core from early on and that in practice nowadays consists largely of mathematical modeling.

The proposals presented to solve the apparent problem of unrealistic assumptions in economics followed two main lines. On the one hand, it has been claimed that despite the apparent falsity of the majority of the assumptions made in economic models, they nevertheless are able to isolate some essential features of economic behavior. On the other hand, those who are not willing to grant such a strong ontological status to economic theories and their parts have invoked the predictive value and understanding-bearing nature of economic models. From this perspective they could be viewed as handy instruments that nonetheless also give some insight into the workings of economics. The crucial difference between these proposals, I suggest, lies in whether or not they subscribe to representationalist realism concerning at least some parts of economic models. However, as I have attempted to show, the representationalist solutions to the problem of the epistemic value of modeling tend to fail both in principle and in practice. Thus, I have proposed an alternative productive approach to modeling. I have also discussed its relationship to fiction, thereby aiming to explain the continuing reliance of economists on the understandability and credibility of economic models, however far removed from the economic reality they may seem to be.

Finally, it should be mentioned that the assumptions issue in economics has never been confined to the philosophical and methodological question of how to justify the practice of idealization and abstraction. More often than not, the target of attack has not been idealization per se, but rather the very content of the idealized assumptions. The constructive and computational strategies that are intended to render models tractable and productive also furnish them with initial built-in justification (Boumans, 1999). For instance, rationality postulates combined with equilibrium concepts that are part and parcel of the majority of economic models are not usually questioned by mainstream economists because of their value as "solution concepts" that facilitate the derivation of the model results. This built-in justification is partly responsible for the built-in inertia visible in modeling[16]—and it is often from this very inertia that the frustration concerning the basic assumptions of economics flows.

NOTES

1. The shift from classical to neoclassical economics happened during the 1870s and 1880s as a result of the so-called Marginal revolution. Neoclassical economics focuses on the determination of prices, outputs, and income through supply and demand in markets.
2. Machlup's numerous writings on economic methodology over the years have appeared in Machlup (1978).

3. Mäki joins several other philosophers of economics in pointing out that instead of worrying about the unrealisticness of the assumptions per se one should study the different kinds of assumptions and their functioning in the economic theory (see, e.g., Musgrave, 1981; Hausman, 1992; Alexandrova, 2006).

4. Morrison (this volume p. 112) pays explicit attention to this problem, too, distinguishing between idealizations, fictions, and abstractions, the last of which introduce a special kind of mathematical formalism "that is not amenable to correction and is necessary for explanation/prediction of the target system."

5. For instance, the rationality postulates do not provide any measurable invariant foundation for economic science even though they constitute its seemingly irrefutable core.

6. In this the philosophy of economics resembles the other fields in the philosophy of science in which models have become a hotly discussed topic—which in turn reflects the growing importance of modeling and simulation in current scientific practice.

7. This indirectness creates the familiar problematic of representationalism. According to representationalism, knowledge is conceived of as an assemblage of representations that reproduce accurately, i.e., stand truthfully for, what is outside the mind (or, after the "linguistic turn," outside linguistic description—or other external representations) (see, e.g., Rorty, 1980, pp. 3–6). The essential difficulty with this theory is that the mind "supposes" that its ideas represent something else but it has no access to this something else except via another idea or representation. One way around this problem is to submit that there are privileged representations or privileged layers in actual representations that capture the world as it is. Structural realism is one such attempt.

8. Mäki (in press) adopts slightly different terminology, characterizing the representative aspect of models in terms of their being "surrogate systems" as opposed to being "substitute systems," which according to him are only studied for their own sake thus substituting for any interest in real systems. As for resemblance, Mäki requires only "resemblance-related issues to arise", which nevertheless "presupposes that the model has the *capacity to resemble* its target."

9. It is well worth noting that the model's being an independent object does not strip it of intentionality. It is intentionally constructed to study certain questions and phenomena whatever its consequent uses are.

10. On the importance of external scaffolding in science, see Giere (2002).

11. Barberousse and Ludwig (this volume) make a related point about the Ehrenfest model—that it is free of the mathematical complexities of statistical mechanical models and is thus more enlightening in terms of which physical properties might explain thermalization.

12. To be sure, Lucas uses the word "artificial" instead of "artifactual." Yet the word "artificial" refers to something that is made by human work, being contrived, pretended, affected or conventional as opposed to natural. It has origins in the Latin word *artificium*, which is based on *ars* "art" and *facere* "make." I prefer to talk about "artifactual" instead of "artificial" because it more clearly refers to the actual modeling process and the tools it makes use of.

13. On the Robert Lucas, Jr., "program" in modeling business cycles, see Boumans (2006).

14. This conforms to how Rouse (this volume p. 45) defines laboratory fictions. According to him, fiction has "sufficient self-enclosure and internal

complexity to constitute a situation whose relevant features can be identified by their mutual interrelations."

15. It should be noted, however, that much as in the case of models, literary fictions, despite their independence, also "amalgamate strata of diverse geological origins" (Pavel, 1986, p. 71). This phenomenon, called intertextuality in the context of literary theory, is one of the very topics of interest to many postmodern authors such as Umberto Eco.

16. Cf. Godfrey-Smith (2006).

Part VI
Fictions and Realism

13 Fictions, Fictionalization, and Truth in Science[*]

Paul Teller

1. INTRODUCTION

We look to science to give us not just accurate predictions but explanation and understanding of what things are and how they work. But how can science do that when and insofar as it frequently appeals to things that don't exist, such as point particles and continuous fluids? Physics, especially, almost always works in terms of distortions, in terms of things that some want to think of as fictional. Does physics, and much of the rest of science, provide us with no more than "useful fictions," useful in providing accurate predictions, but otherwise no more than the product of the imagination?

The idea that science often purveys no more than fictional accounts is very misleading. I will show how the obvious use in science of elements that might be thought of as fictional does not compromise the ways in which science provides broadly veridical accounts of the world.

To understand how science can give us real knowledge of the world even when its accounts dabble in fiction I will appeal to the idea of fictionalization, sketching the notion first generally and then in application to our current problem. To set us up for an effective application I will need first to review some general points about the relation of ideas of being accurate and being true and will then work in terms of the umbrella term "veridical." That will position us to understand how science accomplishes veridical accounts through the use of models. Once we have the modeling lens in place we will better be able to understand the strategy of using

*This chapter is an outgrowth of a workshop, "Scientific Representation: Idealisations, Fictions, Reasons," that took place at Madrid's Complutense University from February 2 to February 4, 2006, organized by Mauricio Suárez. I take myself here to be presenting ideas many of which I learned at the workshop, especially from Eric Winsberg, but also from everyone participating. I hope that I have been able usefully to amplify with some further observations of my own. Readers should especially see Winsburg's and Giere's contributions to this volume, which reach conclusions similar to those presented here with expositions that complement my own.

approximations and idealizations, which will finally apply to show us clearly how the use in science of fictionalization does and does not result in accounts that merit the epithet, "fictional."

2. REPRESENTATION, ACCURACY, AND TRUTH

We need to get our bearings about what we expect from successful accounts in science. The bottom line is that we want representational success in the sense of having representations that represent things as they are. But we are used to thinking in two apparently very different ways about how such representational success can arise. A map may provide an extremely accurate representation of those aspects of a situation in which we are interested. Though a map never depicts everything, nor anything with complete accuracy, when it gets the things that we care about accurately enough to meet our needs, it counts for us as representing things as they are. We contrast such representational success of maps and like analog representations with the idea of truth. Truth is an evaluative term that we apply to sentences, statements, or propositions—I will work with statements for the sake of definiteness. We can understand what we need about the idea of truth in terms of the simplest possible case: A statement that attributes a property to an object is true just in case the object to which the statement refers has the property that the statement attributes. For example, the statement that the earth has a circumference of 40,000 kilometers is true just in case the circumference of the earth is 40,000 kilometers.

The point I want to underscore with this simple example, and a point usually neglected in other discussions, is that we accept such a statement as true when it is "true enough," relative to our present needs and interests, even when it is not true precisely. Generally we accept the statement that the circumference of the earth is 40,000 kilometers as true, even though more accurately the circumference of the earth is 40,075.16 kilometers around the equator. Most wave such inaccuracies aside, thinking that we innocently accept as true a statement that is only true enough because the not quite completely accurate statement functions for us as a stand-in for a completely exact statement, in our example, of the earth's circumference. But of course there is no such thing. For a variety of reasons the earth's circumference is always changing slightly, and even if we try to refine to the exact circumference at a time, which is not what the original statement intended, there are no such things as exact distances: Just what measurement would give the "true" circumference of the earth? Even the fantasy of an ultimately exact measurement is undone by quantum mechanics.

There are various other fixes to which purveyors of exact truth will appeal. I argue elsewhere (Teller, 2008, in preparation) that none of these fixes will work and that we need substantially to rethink how we think about truth. But for present purposes all we will need is that in fact we accept as true a statement such as that the circumference of the earth is 40,000 kilometers when the statement is true enough relative to our present interests, however the wrangling about the finer details might play out. Our example provides a paradigm for what is involved in accepting a statement as true, and so in taking the statement to represent things as they are.

There are two ways in which we can look at such examples. We might understand the statement as that the circumference of the earth is 40,000 kilometers *precisely*, knowing that this is, strictly speaking, false, but we accept it as true insofar as it is true enough, that is, accurate enough for present interests. We accept the false, precise statement as true when no harm will be done in treating the situation as if the circumference of the earth really were 40,000 kilometers precisely. Or, we might understand the statement as that the circumference of the earth is 40,000 kilometers *close enough*. I take it that this imprecise statement counts as true—when?—exactly in the circumstances in which the precise semantic "alter ego" counts as true enough.

The qualification "for present interests" is essential. The statement that the circumference of the earth is 40,000 kilometers seems susceptible to a natural measure of approximate truth. But in more complicated examples there will be different respects in which the approximation may diverge from the reality. Consider a surface that (like all real surfaces) is not quite geometrically flat. But the deviation from geometrically flat may occur in different ways: Maximum deviation from an average? Maximum slope of the surface relative to an average? Different interests will require different degrees of accuracy in these and in other respects. Even in our first example, the circumference of the earth around the equator is more nearly 40,075.16 kilometers but around the poles it is 40,008 kilometers.

In every case a statement that counts as true enough, or as accurate enough, will be incomplete until we specify both the respects of interest and then a standard of deviation that must be met. I will generally mark such relativity of statements and of other kinds of representations with the phrase "for present interests."[1]

The upshot is that, whatever might be said about in principle refinability, the statements that we appropriately accept as true do represent things as they are, but much more in the manner achieved by maps and other such analog representations when we appropriately accept them as accurate. I will use the term *veridical* as the umbrella success term when, with respect to present interest, a representation succeeds in representing things as they are, in the way achieved by an accurate map, a true (enough) statement, and other sorts of accurate but not completely exact representations.

For a glance ahead, clearly a veridical representation will never count as a fictional representation, or, a little more carefully, the respects in which a representation counts as veridical will never be respects in which the representation counts as fictional. Complete precision and accuracy are not required to count as nonfictional.[2]

So far, so good. But when we describe an extended object as a point particle or water as a continuous medium we are not providing veridical representations. Agreed. Such representations fairly count as fictional. But it does not follow that when we embed such a contained fiction in some larger representation, the containing representation is thereby a fiction or is fictional. The balance of this chapter shows that not only does such a conclusion not follow, but in cases of interest such a conclusion is mistaken.

3. MODELS AND REPRESENTATION IN SCIENCE

We are interested in the question of whether, or to what extent, representation in science is fictional. To examine this question we need to know very generally the form in which science provides representations. It is now widely recognized that most, if not all, of the representations of the real world that science provides come in the form of models.[3] For our present topic we need a sketch of what models are and how they represent.

We begin with a few examples. One may provide a model of the solar system by building a physical model with suspended balls representing the sun and planets. Or one can provide an abstract object that exhibits the same relevant structure. A model of a chromosome may again take the form of a physical object, such as Watson and Crick's original, or be an abstract object that achieves the same goal of representing the target by similarity of structure. Generally the models that we use in science are abstract, and often can only be abstract, as when we model water as a continuous incompressible medium governed by Euler's equations.

A model is an object, concrete or abstract, that bears some relevant similarity to what we want to represent. The similarity itself does not make the object a representation; it is we, the model users, who press the similarity into representational service. So, to represent a target and its features we use a representing object together with (explicit or often tacit) statements of the relevant similarities between the object and the representational target. We usually call the representing object itself the model, but it is a model of the representational target only in virtue of our treating the relevant similarities as representational.

Models usually have a variety of components or parts. For example, in a model of the solar system we might represent the sun as a point particle. For a systematic way of talking about representational components of a model I will use a term for the object represented surrounded by quotation marks. So, in the first example we could use either the expression "'the sun'" or "'the most massive point particle'" to refer to the part of the model that we use to represent the sun. (Strictly speaking I should have enclosed 'massive' in quotation marks also—"'massive'"—because a "point particle" in an abstract model, strictly speaking, has no mass: The "point particle", as a component of an abstract object, is itself abstract, and so not the sort of thing that has one or another mass. Rather, the "point particle" is characterized abstractly as to the represented quantity of mass. But within the outermost quotation marks, this will be understood.) Early discussions of models, especially discussions using the idea of a model as developed in the formal discipline of model theory, often worked in terms of claimed isomorphism between the representing object and the representational target. But models in science almost always work in terms of similarities that fall short of exact isomorphism. So scientific models fall naturally into the category of representations exemplified by maps that work in terms of similarities in respects and degrees of interest.

4. APPROXIMATION AND IDEALIZATION

Approximation and idealization provide basic tools for model building. These are open-ended ideas that get used loosely and in overlapping ways. The following characterization corresponds well enough to common usage and will suffice for our needs. One generally speaks of approximation for a use of a mathematical expression or quantity when there has been substitution of a simpler expression that is close enough to the correct expression not to spoil the intended application. For example, one speaks of approximation when one redescribes an ellipse with a small eccentricity as a circle, or when one uses an angle instead of the sine of the angle, which provides a good approximation for small angles. In loose analogy one also speaks of approximation for nonmathematical expressions, especially when these are susceptible to some mathematical characterization. So one might provide a simple drawing, saying of it that it provides the approximate look of some city skyline; or, pointing at a color sample, one might say that the indicated color is approximately the color of one's car.

On the other hand, generally an idealization involves some radical misdescription in a way that simplifies or otherwise facilitates an account. "Point particles," "continuous media," and "perfectly rigid bars" provide examples of idealizations. Often one speaks of idealization when leaving out some parameter, causal feature, or other factor in representing a situation, as when leaving out the force of friction in characterizing a "frictionless plane" or leaving out the force of gravity between two objects when that force is very small compared to an electrostatic force.

The distinction between approximation and idealization is rough and ready. Often what one calls an idealization could also be described as an approximation, especially when one has substituted a limit value for a quantity, as in the two last examples: Neglecting the force of friction or gravitation is to characterize the force as having value zero. Generally, leaving out some quantity or factor can often be called either an approximation or an idealization.

5. TWO SENSES OF "IDEALIZATION"

Idealization constitutes an entry point for discussion of fiction in models. I am going to suggest that misunderstanding about fiction arises when one conflates two different ways of thinking about idealization.

When one learns a science, one learns various general strategies for building models, a great many of which we can think of as *idealizing strategies*. For example, when one is concerned with nonrotational motion of an (extended) object, very often one can simplifying the problem by redescribing the object as a point particle located at the object's center of mass. When one is interested in the fluid properties of a liquid, as opposed to its properties as a solvent and agent of dispersion, one can do very well by describing the liquid as a continuous medium. The idea on which I want to focus is that when one uses such an idealizing strategy, the strategy instructs us to construct some specific idealization that is to be used as a *part* of a model. Thus, some larger constructed model will include, as a part, the idealized object, a "point particle" that functions to represent some extended object as a point particle; or, some larger constructed model will include, as a part, a "continuous medium" that functions to represent some liquid as a continuous medium. I will call such an idealized component of a model a *component idealization*.

Now let's shift focus from the idealized part of a model to the model as a whole. If the model is successful, in respects other than the serious distortion involved in the component idealization, the model will be accurate in respects in which we are interested and to the degree that we require. Thus, except for the (one or more) component idealizations, a successful model will accurately represent the world, and in so doing will count as veridical with respect to the aspects in which it is accurate. But people have the practice of using the expression "idealization" for such models as a whole. That is, even when a model is successful in representing the things we are interested in at the moment, when such a model has as a part a component idealization the practice is, as it were by courtesy, to use the term "idealization" to characterize the model as a whole. When it is the model as a whole that gets called "an idealization" on the strength of the fact that the model includes some component idealization, I will use the term *containing idealization* for the model as a whole.

6. TRUTH AND FICTION

In preparing to address the question of whether, or when, models qualify as fiction, we have examined the first half of the equation: models. It is time now to turn to the second half: fiction.

We think of truth and fiction as exclusive contraries. At best, so saying caricatures a subtle and complex interplay. In the first place, every bit of fiction includes or at least presupposes some truths about the world. A novel must presuppose some general facts about people, or at least about the nature of life if it will be a novel about aliens. To be fictional an account must in some way misdescribe some real or some real kind of thing. If a fictional account did not connect in some veridical way with things with which we are familiar, the fiction would have no meaning for us. There is no such thing as a "fictional" counterpart of completely abstract painting.

We will be interested in the complementary fact that discourses that we correctly accept as true *can* have a little bit of fiction: "Ladies and gentlemen: the story you are about to hear is true. Only the names have been changed to protect the innocent." This opening line from the old-time radio serial *Dragnet* makes perfectly good sense. There are many serious life examples of the same thing. Many medical case studies must change details in the description of the subject to preserve the subject's anonymity. This obligation may require judicious misdescription, for example, in psychoanalytic case studies where many personal details will be relevant to the medical point in question. The case writer must carefully alter details that are not relevant while preserving the ones that are, so that the report will convey a true account of the medical facts while hiding the subject's identity.

Such cases are instances of the process of fictionalization, a process that can result in an account that is, in respects in which we are interested, still veridical. We need to say more about fictionalization and its relation to fiction.

7. FICTION, FICTIONS, AND FICTIONALIZATION

A fiction is not a mistake. A mistake is a misrepresentation that occurs unwittingly, where the author of the representation is unaware of the departure from veracity. A fictional discourse or other sort of representation also involves departure from a veridical account, but to qualify as fictional the departure must be purposeful, must involve some purpose, intent, or objective that one self-consciously seeks to achieve by means of constructing the fictional elements of the representation. In familiar cases the purpose will be to entertain and sometimes to deceive. In science the purpose is almost always to facilitate representation that in other respects is veridical.

In the first instance, to label a discourse as fiction is to place it in a genre of writing characterized as products of the imagination as opposed to the reporting of facts: We speak of "works of fiction." In this sense, the words "fiction" and "fictional" apply to the discourse as a whole. We think of the whole as departing from giving a veridical account of any part of the world. And we don't, on the whole, think of works that are fictional in this sense as being embedded or embeddable in some larger discourse.

Although "fiction" and "fictional" are in this sense usually applied to writing, the terms can be applied to other kinds of representations, such as a fictional map. I will call fiction in this sense *fiction in the inclusive sense*.

There is another use of "fiction" that we hear more readily when used in the plural: fictions. A fiction in this sense is some fabulated event, state of affairs, object, or person. A fiction in this sense is a representation that purports to represent an object or person that does not really exist (an *object fiction*) or purports to represent a state of affairs or event that may concern a real object or person but that does not or did not really occur (*an event or state of affairs fiction*). Santa Claus is an object fiction. We describe event fictions with comments such as, "Smith is a member of the soccer team, but the description of him as having scored with a bicycle kick is a complete fiction." I will use the expression *component fiction* for object, event, and state of affairs fictions.

Unlike fiction in the inclusive sense, a component fiction can be put to use in some larger discourse, in particular a larger discourse that may otherwise be veridical.[4] To illustrate: "Mary's description of the soccer game was very accurate, except for Smith's bicycle kick goal, which was a complete fiction." Again: " Some of the presents that children get at Christmas are delivered by Santa Claus." In this second example, both Santa Claus and his claimed delivery of presents are fictions embedded in the correct description of children getting presents at Christmas.

I am interested in the idea of component fictions because of the role it plays in the process of fictionalization. One way to think about the process is as starting with a true story and then replacing smaller elements of the account with corresponding fictions that, however, retain many of the relations among events that are chronicled in the story.[5] When one retells extensive parts of a story in this way the result is a fiction in the inclusive sense, something that as a whole counts as fictional. We see examples of such fictionalization in the biographical novels of Irving Stone and in biographical films ("biopics") such as *Schindler's List*, *Gandhi*, *A Beautiful Mind*, and many others.[6]

I want to contrast the last kind of case with ones in which we want to "tell a true story," but in a context in which we must meet some constraints in the telling. These are cases in which one wants to preserve a veridical discourse, or at least the veracity of important parts of it, but one can only do so while satisfying constraints that force some compromises. Such cases also proceed by substituting fictions for truths within the story, but only when and insofar as required by the constraints. As in the introductory examples given earlier, one may be constrained to change the names of people mentioned "to protect the innocent." Or, in giving a veridical psychoanalytic case history, one must change many personal details in order to preserve the anonymity of the subject. Something relevantly analogous occurs when, for example, in order to make calculations manageable one must redescribe an extended object as a point particle. Often these adjustments occur without explicit comment. We often presuppose sophistication on the part of the audience as to what to take literally and what as ancillary and nonliteral, or fictional.

In cases like these it would be very misleading to say that the discourse or other representation has been turned into something that as a whole counts as fictional, that as a whole counts as a fiction in the inclusive sense.

Let's refer to fictionalization of this second sort as *veridical fictionalization*. The objective is accurately to communicate some salient content. In order to meet a constraint on providing such a representation, we fictionalize some elements that are not relevant to this salient content. In the process the accuracy of the salient content may be degraded to some degree. But as long as the salient material is still represented accurately enough to meet current interests and needs, we still appropriately characterize the representation as a whole as veridical, even, as in the case of *Dragnet*, as true. And this even though the representation includes, as parts, fictions in the component sense.

Component fictions do not generally turn a larger veridical representation itself into a fiction or make it, as a whole, fictional.

8. MODELS, FICTIONS, AND ACCURATE REPRESENTATION

What goes for descriptions with fictional elements likewise goes for models. A model may have fictionalized elements that in themselves radically misrepresent but that, nonetheless, facilitate the accurate representational function of other aspects of the model. The aim is veridical representation. But there are heavy practical constraints on calculation and other manipulation of parts of a model. We want to get out numbers. We want to see how the "moving parts" work so that we can understand the mechanism in question. A crucial part of the art of science is to figure out how to make idealizing simplifications, distorting only things that, for the context in question, are relatively unimportant but that otherwise preserve accuracy, at least sufficient for present purposes. By fictitiously characterizing a real extended object as a point particle we have to apply Newton's laws only to the point, not to the whole, often very irregularly shaped, body. Only by fictitiously characterizing water as a continuous medium can we characterize its fluid behavior in terms of hydrodynamical differential equations. So insertion of these component fictions plays a most positive role in achieving representations that are both veridical and manageable. And this notwithstanding the fact that using the point particle idealization completely misdescribes the shape of the object, and using the idealization of a continuous medium completely misdescribes the fine texture properties of water, which, in turn, throws a monkey wrench into efforts to understand things such as dispersion and solvent properties.[7]

The common practice is to think of idealizations such as point particles and continuous fluids as what we are calling object fictions. There are no such things. This is misleading. Abstract elements in containing models, such as "point particles" and "continuous fluids" (remember the use of quotes to indicate abstract elements of models), function to represent *real* objects such as extended objects and bodies of water. But these model components accomplish such representation in a very idealized way. A real extended object is fictionally described as having no extension. A real body of water is fictionally described as being a continuous fluid. Such cases constitute fictional description of real objects. So such cases should be thought of, not as object fictions, but as state of affairs fictions, as fictional characterization of states of affairs of real objects. These fictional characterizations are then used as component fictions to represent the real trajectories of real objects, the real fluid behavior of real bodies of water. When a model is successful, such representations of the motion of an object, of the fluid behavior of a body of water, are accurate enough to count as veridical and so anything but fictional.

There are also cases of component object fictions that unequivocally involve nonexistent entities. The method of image charges illustrates this kind of case. Suppose that we have a metal plate and a fixed positive electric point charge a short distance, d, from the plate. What is the induced distribution of charge on the plate? Imagine a different situation absent the metal plate but with a negative charge of the same magnitude as the original positive charge the same distance, d, on the other side of where the plate had been. With relative ease one can calculate the electric field resulting from the two point charges. The plane on which the plate had been located is an equipotential plane. So if we reinsert the plate and remove the image charge, nothing will have changed on the side of the plate with the original charge. Finally, from the electric field, which we now know, we can use Gauss's theorem easily to calculate the distribution of charge on the plate. The image charge functions as an object fiction in this example.[8] Eric Winsberg's example of silogens provides another example.[9]

Earlier we differentiated between component and inclusive idealizations. Generally speaking, component idealizations are component fictions, fictional representations of real objects, or, occasionally, object fictions. In the case of idealization we engage in the practice of calling the whole model an idealization—an idealization in the containing sense—when the model includes one or more component idealizations. There is no harm in this. But then we are tempted by the obvious parallelism. Component idealizations are component fictions. We call a model that includes a component idealization itself an idealization—in the containing sense. Then we are tempted also to call the containing model a fiction, if only by courtesy, and in a containing, and so in what I called the inclusive sense of fiction. This is what we appear usually to do. Any such conclusion is misleading in the extreme.

When a model is successful it is, in the respects that we care about, veridical—and that, as we have seen, often on the strength of component fictions. Component fictions play the role of meeting representational constraints in otherwise veridical representations, as in the case of psychoanalytic case reports, the constraint of preserving the anonymity of the subject, and the case of modeling the linear motion of an extended object, the constraint of making the calculations manageable. But if we are seduced by the forgoing parallelism we are stuck with the apparent conclusion that such successful models are, as a whole, fictional or "merely useful fictions," which contradicts the fact that, in the case of successful models, the models are in relevant respects veridical.

So we must insist that a successful model that makes use of component fictions is not thereby itself fictional. In very special cases we may appropriately call an entire model a fiction, as for example in thought experiments and so-called "toy models," such as φ^4 models in quantum field theory that use an utterly unrealistic potential.[10] Practioners sometimes use such "toy models," not to describe real systems but to illustrate how the relevant mathematics works.[11] But successful models are not themselves fictional just because they use fictional elements.[12]

Thus far I have presented the idea of a veridical but in some way fictionalized version of an exact account by appeal to the notion of idealization. But because idealization and approximation are not clearly distinguished, we should expect that at least some cases of approximation should also count as fictionalizations. What counts here is not the term "approximation" or "idealization" but the pattern of component fictions playing a role in otherwise veridical representations. Relative to possible interests in distorted aspects of a component idealization or approximation, the idealization or approximation will count as a contained fiction. But relative to interests for which the containing model still produces acceptably accurate descriptions and explanations, the model will count as veridical.

9. SHIFT FROM MISTAKE TO FICTIONAL ELEMENT

Looking at the relation between mistakes and component fictions will further illustrate and apply important parts of our analysis.

Imagine that when Euler developed his hydrodynamical equations everyone actually believed water to be a continuous medium.[13] These equations give an excellent description of many aspects of the fluid properties of water. Then practitioners discovered that water is not a continuous medium— practitioners realized that in characterizing water as a continuous medium they had made a mistake. But they, and we today, continue to use Euler's equations in a wide range of applications.

This example illustrates the way in which what had the status of a mistake can acquire the status of a fiction. After the discovery, characterization of water as a continuous medium is retained for a wide range of models that give excellent results for many aspects of fluid properties, including not just the capacity for accurate predictions but also explanatory understanding of the mechanics of the fluid behavior of a substance such as water. But now that we know that water is not continuous, its characterization as continuous takes on the status of a fictional element in otherwise very accurate models.

Characterization as a mistake or as a fiction function as epistemic categories. As we noted before, a fiction is never a mistake. A mistake is a claim made in the belief that it is true or accurate, although in fact it is false or inaccurate. A fiction is also a description that is false or inaccurate. But it is one that is known as such. When we find that a former mistake-turned-fiction facilitates otherwise veridical representations, we continue to put it to good use in producing veridical accounts of the world.

10. CONCLUSION

Truth and fiction are often taken to be polar opposites. We have seen that their relation is much more subtle. In many ways, fiction can play a vital role in veridical representation of the world.

NOTES

1. For more on this issue see Teller (2001, pp. 401–406).
2. See Giere's contribution to this volume for a discussion of considerations bearing on how we classify works as fiction or as nonfiction.
3. See Teller (2001) for a sketch, defense, and further references.
4. The expression "component fiction" is appropriate because, although such a fiction does not actually have to be a part of some more inclusive representation, such fictions always can be so included.
5. I am not claiming that this is the only or even the most common way in which fictionalization proceeds. Clearly this is one way in which fictionalization proceeds, and a more detailed examination of the process of fictionlization would have to study the question of whether we also fictionalize in other ways.
6. The fictional nature of biopics is underscored by the controversies they provoke, up to and including threats of laws suites. See the Wikipedia entry on Biographical Film at http://en.wikipedia.org/wiki/Biopic.
7. These can be "put in by hand," but in a continuum model they cannot be described in an explanatory way.
8. Ron Giere (this volume) also points out that in working the problem I have just described one works within a model that describes the charge as a point charge and the metal plate as an infinite surface, and that "[i]t is telling that textbooks do not refer to [the positive point charge and infinite metal surface] as "fictional" although they are clearly physically impossible entities."
9. See Winsberg's contribution in this volume.
10. φ^4 theories do have some real-world applications, but they are often used for purely illustrative purposes.

11. A purely fictional novel may provide a good illustration of some psychological or moral point.
12. Cartwright (1983, p. 153) writes that "A model is a work of fiction," but her comparison with historical writing and theatrical presentation of real events (pp. 139–142) would seem clearly to show that she intends a view with a substantial amount in common with what I am presenting here.
13. Historical accuracy is not here to the point. We are interested in the ideas in this example.

14 Why Scientific Models Should Not Be Regarded as Works of Fiction

Ronald N. Giere[1]

1. INTRODUCTION

The usual question in this context is: Are models fictions? I find this way of formulating the question to be misleading. It suggests that both scientific models and works of fiction have intrinsic, essential properties, so that, once these are disclosed, we can easily ascertain whether these two sets of properties are identical or, more realistically, the extent to which they overlap. Given the variety of things that can be used as models as well as the variety of fictional genres, I doubt any useful set of such intrinsic properties exists. I have thus implicitly replaced the usual question by another: Should scientific models be regarded as works of fiction? This way of stating the issue makes clear that the issue is not one of definition but of interpretation. And it suggests that possible answers to the question are to be evaluated in terms of their consequences for understanding the roles of models in the practices of scientists. On this understanding of the question, I think the answer is decidedly negative.

Why would one even think of regarding scientific models as works of fiction? I discern three reasons in the literature. First, scientists themselves sometimes invoke the idea of fictions in their discussions of specific models. So we theorists of science have to make sense of such discussions. Second, many scientific models are physically impossible to realize in the real world. The model of a simple pendulum, for example, contains no representation for friction, so any exact realization would be a perpetual motion machine, which is universally regarded as being physically impossible. How are we to understand the ubiquitous use of such models? Are they themselves, or do they represent, fictional entities? Third, regarding scientific models as fictional may be part of a general fictionalist understanding of scientific theories. On this view, theoretical models contain or represent fictional entities. Only representations of things that may be experienced in practice can be veridical.

In the end I will reject each of these reasons for regarding scientific models as works of fiction. But first I want sharply to distinguish between the *ontology* of scientific models and their *function* in the practice of science.[2]

Theoretical models and works of fiction are, I will argue, ontologically on a par. It is their differing functions in practice that makes it inappropriate to regard scientific models as works of fiction.

2. ARE FICTIONS MODELS? THE ONTOLOGY OF FICTIONS AND MODELS

Scientific models seem to be ontologically quite diverse, including physical scale models, diagrams, and abstract (or theoretical) structures. There are several ways of dealing with this diversity. One way is to take physical models, diagrams, and so on as being ontologically unproblematic and concentrate on the more problematic abstract models. A second, more radical, solution is to regard physical models, diagrams, and so on, as resources for partially characterizing abstract models. So, all scientific models are regarded as being abstract, or at least having abstract counterparts. For the present inquiry it does not matter which of these alternatives one chooses because it seems to be only abstract models for which there is any question about their possible status as works of fiction.

One way of approaching questions about the ontology of models and fictions is to ask the reverse of the standard question about relationships among models and fictions. Are fictions models? I would, of course, rephrase this as the question of whether in creating a work of fiction one should be regarded as also creating models. Here I think the answer is decidedly positive. Consider a quintessential work of fiction, such as the novel *War and Peace*. I claim that Tolstoy can be regarded as having created an elaborate possible world, which is to say, a model (perhaps consisting of many less elaborate models) of a possible world.[3] This model in fact resembles the real world in many respects. It speaks of such actual things as Napoleon's invasion of Russia, life and death, love and loss. It was, indeed, intended to represent such things. To the best of my knowledge, this is a model that could have been realized in the real world. It is what Peter Godfrey-Smith (2006) has called an "imagined concrete thing." In fact, it has been realized several times in films. That the original work describes a model is shown by the fact that the film realizations necessarily add details, such as some features of the main characters. But the films, too, like the novel, are just models of a possible world characterized in terms of spoken words and pictures. Of course, not all works of fiction describe things that could be realized in the actual world. Mark Twain's *A Connecticut Yankee in King Arthur's Court*, which invokes traveling backward in time, is a well-known example of such a work.

It is widely assumed that a work of fiction is a creation of human imagination. I think the same is true of scientific models. So, ontologically, scientific models and works of fiction are on a par. They are both imaginary constructs. It is also relatively uncontroversial that the ability to create imaginary objects is facilitated by language. This is very clear in the fundamental case of planning and decision making. In response to a given situation, one may imagine and describe possible courses of action and likely outcomes of these possible actions. In an ideal case, only one of these possible actions is carried out, leaving the rest as merely imagined possibilities. Of course, imagined scenarios vastly underdetermine, and indeed are typically contradicted by, the actual course of events.[4]

The connection between producing scientific models and producing works of fiction must be deeper still. Surely the imaginative processes at work in producing a scientific model are similar to those invoked in producing a work of fiction. There are many ways one might study these processes. For example, one might have expected that the contemporary cognitive sciences could shed some light on these processes. That seems not to be the case.[5] There does, however, seem to be a large literature on the general nature of fiction and fictional entities among literary theorists and some philosophers.[6] One can also study authors who produce works of fiction, as well as scientists who have constructed significant scientific models. Many works of biography by historians and historians of science are explicitly concerned with what is commonly called "the creative process."[7] The question for me, however, is whether we, as philosophers of science interested in understanding the workings of modern science, *need* a deeper understanding of imaginative processes and of the objects produced by these processes.

To take a case close to home, when I call theoretical models "abstract," I mean mainly that they are "not concrete," which is the core meaning given in dictionaries. Objects in the world, including scale models, are "concrete." One can, of course, inquire further into what it might mean to call theoretical models "abstract." Are they, for example, similar to mathematical models? Does the fact that abstract models do not exist in actual space and time imply that we cannot talk about the period of a simple pendulum? Do the terms "length" or "period" mean the same thing when applied to the model of a simple pendulum as when applied to a concrete pendulum?[8] If, ontologically, models are imaginary objects in the way objects of fiction are imaginary objects, the answers to these questions are negative. We surely would not want to say a character in *War and Peace* is a mathematical object, nor that such a character could not be described as traveling from one place to another, nor that the word "war" means something different in the novel than in real life.

More positively, my view is similar to Cartwright's (1999a). Some things she calls "laws" I call "theoretical principles," which characterize (and are true of) highly abstract models, which, however, cannot be used to represent anything in particular. For that one needs more specific models that embody the principles. But even a fully specified model that is intended to represent a particular concrete system is still an abstract object. I remain to be convinced that we need say much more than this to get on with the job of investigating the functions of models in science.[9]

3. FICTION, NONFICTION, AND SCIENCE FICTION: THE FUNCTION OF SCIENTIFIC MODELS AND WORKS OF FICTION

In spite of sharing an ontology as imagined objects, scientific models and works of fiction function in different cultural worlds. One indication of this difference is that although works of fiction are typically a product of a single author's imagination, scientific models are typically the product of a collective effort. Scientists share preliminary descriptions of their models with colleagues near and far, and this sharing often leads to smaller or larger changes in the descriptions. The descriptions, then, are from the beginning intended to be public objects. Of course, authors of fiction may share their manuscripts with family and colleagues, but this is not part of the ethos of producing fiction. An author of fiction would not be professionally criticized for delivering an otherwise unread manuscript to an editor. Scientists who keep everything to themselves before submitting a manuscript for publication are regarded as peculiar and may be criticized for being excessively secretive.

Turning to a more public arena, the table of contents of the weekly book review section of the *New York Times* has two main categories: fiction and nonfiction. Although reviewers often point out mistakes in works of nonfiction, that is not taken as a ground for moving these works into the fiction category. It does, however, sometimes happen that a work marketed as nonfiction turns out to describe events many or even most of which are later discovered never to have happened. Trying to pass off a work of fiction as nonfiction may destroy an author's career. There have even been cases in which an author's career was derailed for trying to pass off fictional composites of real people as a single real person. Thus, although the boundary between fiction and nonfiction is not sharp, it is well understood.

This applies to science as well. Works popularizing new scientific theories are listed as nonfiction. Works of science fiction are, of course, listed as fiction. Even books describing 11-dimensional vibrating membranes or multiple universes, which sound to many ears as being "stranger than fiction," are classified as nonfiction as long as they are written by recognized scientific authorities and intended as at least possible scientific descriptions of the world.

In fact, regarding claims about the fit of scientific models to the world as fictional destroys the well-regarded distinction between science and science fiction. Thus, imagining light traveling in totally empty space, which we now think does not exist anywhere, can be part of good science. Imagining the spaceship *Enterprise* traveling at "warp speed" (faster than light) is science fiction. The general view that to create a scientific model is to create a work of fiction obliterates this well-founded distinction.

There also seems to be no one primary function for works of fiction. At their most exalted, such as *War and Peace*, they provide insights into "the human condition," albeit in a culturally specific time and place. But there are whole genres of fiction, such as romance/fantasy, whose main function seems to be simply to entertain. Then, of course, there is pornography.

Scientific models also serve many purposes.[10] One purpose, however, stands out from the rest. Scientific models typically function as a means for *representing* aspects of the world. Representational virtues are many, such as scope, accuracy, precision, and detail. These virtues may sometimes conflict; for example, one must often trade off scope against detail. Still, a failure appropriately to represent its target is a ground for criticism, even rejection, of a scientific model. Remember the many models that were proposed and rejected in the race for the double helix because they failed adequately to represent the structure of DNA molecules. In the realm of fantasy, such criticisms are not appropriate. It is no criticism of the Harry Potter novels that there is no community of genuine wizards. Nor is it a criticism of *War and Peace* that its main characters did not exist.

4. SCIENTISTS' FICTIONS

One motivation for taking seriously a connection between scientific models and works of fiction is that scientists themselves sometimes explicitly invoke just such a connection. A brief look at some examples of this phenomenon suggests, however, that there is no implication that scientific models themselves should be regarded as works of fiction.

Paul Teller (this volume) reminds us of the following textbook example of the method of image charges. Imagine an infinite metal plate with a positive point charge located a distance, d, from the plate. The problem for the student is to determine the induced charge distribution on the plate using known principles of electrostatics. Applying the method of image charges, one replaces the original model with a model in which the infinite metal plate is replaced by a "fictional" negative charge placed symmetrically on the other side of where the surface had been in the original model. The solution to the problem using the new model, in full accord with electrostatic theory, is exactly the same as if one had solved the mathematically more difficult problem using the initially suggested model.

What is meant by calling the negative charge in the second model "fictional"? As a component of a model, the image charge in the second model is no more and no less fictional than the positive point charge and infinite metal surface in the original model. It is telling that textbooks do not refer to the latter as "fictional" although they are clearly physically impossible entities. My analysis of the situation is that the original model is understood to be an idealized representation of a concrete system. The concrete system would only have counterparts to the original positive charge and conducting surface. Relative to this suggested concrete system, the negative charge in the second model is called "fictional" because it would have no counterpart in the assumed concrete system. On this understanding of the situation, there is no basis for calling either model as a whole a work of fiction.

Eric Winsberg (this volume) describes an advanced contemporary case of a model with a "fictional" component. The problem is to model a growing fracture in a micrometer-sized piece of silicon. At the point of the fracture, the best models are based on quantum mechanical (QM) principles. A little further out, QM models become computationally intractable. So in this region one employs models using principles from classical molecular dynamics (MD). Further out, MD models become computationally intractable, so the outer region is modeled using principles from classical continuum mechanics (CM). There is a problem, however, in that there is no computationally tractable way of modeling the transition from the QM region to the MD region. This problem is solved by introducing into the model molecules of a fictional type, dubbed "silogens," that have some properties of silicon and other properties of hydrogen. In the resulting model, silogens are placed on the boundary between the QM and MD regions, and make modeling the transition computationally tractable. The payoff is that the resulting model both is computationally tractable and can be used as a scientifically and technologically adequate representation of the whole concrete system.

As in the case of the method of image charges, silogens are not called "fictional" because they are elements of an idealized model. That is true also of the elements of the models based on quantum mechanics, molecular dynamics, and continuum mechanics. Rather, while the latter models are taken to provide a useful fit to real features of the target system (micrometer-sized pieces of silicon), silogens are not intended to model anything in a concrete system, it being known that there are no such things in the real world. Relative to the real world, they are fictions, but this role provides no grounds for regarding the overall model as a work of fiction. This model provides a scientifically adequate representation of real systems.

Not all examples of models for which scientists themselves invoke the notion of fictions involve just a fictional element in a model that as a whole is representationally adequate. Some models are, as a whole, fictional. Maxwell's mechanical model of the supposed ether is a much-discussed example of this sort (Nersessian, 1992; Morrison, this volume). There is probably no single function served by all such models. In the Maxwell case, the fictional mechanical model provided a crucial analogy for the construction of a field theory of electromagnetism. So one can say that its function was heuristic rather than representational. But, as Morrison argues, this is a mere template for understanding its role in Maxwell's thought. What matters is how the analogy functions, and this will vary from case to case. In Maxwell's case, a more informative description of the role of the mechanical analogy is that it provided analogs of relationships among variables in the final electromagnetic theory, but this description still needs to be filled out in more detail. The important point here is that there are no grounds for regarding Maxwell's mechanical model as functioning like a work of fiction.

Finally, the fact that scientists themselves sometimes use the word "fictional" to describe an element of a model, or even a whole model, provides little ground for concluding that these models *function* as works of fiction in anything like the way novels function as works of fiction. My guess is that when scientists describe part of a model as fictional they mean little more than that it has no counterpart in reality. The term "fictional" in these contexts is a term of convenience that has few meta-theoretical implications. It is doubtful that many scientists ever consider the possibility that fictional models might function like literary works of fiction. It is more doubtful that they distinguish between the ontology and the functions of models. In general, scientists are not experts in the meta-theory of their practice. In fact, scientists' meta-theories often derive uncritically from diverse sources such as professional folklore and popularized philosophy of science. They typically have no worked out conceptual scheme in which even to state their meta-theory. Their science does not require such. So knowledgeable outsiders such as historians and philosophers of science may have a better meta-conception of what they are doing than they themselves. One task for historians, philosophers, and sociologists of science is to provide scientists with a better meta-understanding of their own practice.

5. THE ARGUMENT FROM IMPERFECT FIT

A second apparent motivation for associating scientific models with works of fiction is the undisputed fact that most abstract models, particularly those characterized mathematically, cannot possibly exhibit a *perfect* fit to any real system. According to our most fundamental theories, it is physically impossible for the world to include mass points, frictionless motions, perfect vacuums, perfectly flat space–time regions, infinite populations, and so on. But this consideration hardly shows that claims about the *good* fit of models to the world function as fictitious claims. Models incorporating such objects figure prominently in good scientific practice and often provide a very good fit to their intended targets in the real world. It violates scientific practice to regard claims of good fit for these models as functionally fictional claims. Idealizations, abstractions, and approximations, yes.[11] Fictions, no.

In fact, the argument from imperfect fit to a functionally fictional status for models proves far too much. It is a commonplace of cognitive science that all thought and communication involves idealized categories, with stereotypes being a prominent example of this phenomenon. Similarly, for several decades, cognitive scientists have argued that ordinary empirical categories are not strictly binary, but graded from central to peripheral cases (Smith & Medin, 1981). So if idealization and the resulting lack of perfect fit imply that models should be regarded as fictions, most of what everyone thinks and says should be regarded as fictional. Once again, the functional distinction between works of nonfiction and works of fiction is obliterated.

It seems to me that the assimilation of scientific models to works of fiction presupposes an exaggerated conception of nonfiction. On this conception, a genuine work of nonfiction has to provide "the truth, the whole truth, and nothing but the truth." Thus, the realization that scientists are mostly in the business of constructing models that never provide a perfect fit to the world leads to the unwarranted conclusion that scientists are in the business of producing fictional accounts of the world.

It also seems to me that part of what drives the move to fictions is the assumption that scientists must mostly be talking about real systems, so when they ascribe features that we know do not exist to a real system, we are forced to look for some other conception of what they are doing. That they are creating fictional properties is one possible conception. But the problem is much less acute if, instead, we conceive scientists as talking much of the time not about a real system but about an abstract model of a real system.[12] The model can have properties we know the real system does not. Then the question is, as always for models, how similar the model is to the real system of interest, which lacks counterparts for some elements of the model. In addition, utilizing similarity as the relation between models and the world helps to solve the problem raised by models not exactly matching the world due to abstraction, approximation and idealization.

6. NEO-FICTIONALISM

A third and even deeper philosophical motivation behind the movement to regard scientific models as works of fiction is its apparent support for what I am here calling "neo-fictionalism." It might also be called "neo-instrumentalism," or, if a more positive sounding label is desired, "pragmatism." One gets more than hints of this motivation in the locus classicus for current discussions of fictionalism, Arthur Fine's 1993 paper, "Fictionalism." This paper is a sympathetic commentary on one of the most extensive works on fictionalism ever written, Hans Vaihinger's *The Philosophy of 'As If'* (1924). As Fine points out, Vaihinger, who saw fictions everywhere, nevertheless contrasted "fictions" (actually "semi-fictions") with what he called scientific "hypotheses," which he thought could be straightforwardly true or false. Thus, although Vaihinger rightly did not attempt to provide a sharp demarcation between "fictions" and "hypotheses," he nevertheless maintained a robust sense of the difference between fiction and nonfiction, at least for science.

Fine is in clear agreement with Vaihinger on the point that "Vaihinger regards the inference from utility to reality as fundamentally incorrect"

(p. 8) Or, as restated a few lines later, he claims that Vaihinger was very concerned to demonstrate "that the inference from scientific success at the instrumental level to the literal truth of the governing principles is thoroughly fallacious." These are claims that Fine himself has often made in his own behalf (Fine, 1986). They are part of his critique of realist interpretations of theoretical scientific claims. So Fine's fictionalism, in the persona of Vaihinger, is being employed in the service of antirealism (or at least nonrealism). This judgment accords with the fact that Fine's paper begins and ends with references to contemporary disputes between realists and antirealists. It accords also with his discussion of the last recent philosopher of science seriously to discuss fictionalism, Israel Scheffler (1963).[13]

Because this is clearly not the place to argue these deeper issues at any length, I will make only one major point. The claim "that the inference from scientific success at the instrumental level to the literal truth of the governing principles is thoroughly fallacious" is itself fallacious. It is so because both the supposed premise and conclusion of the inference are misconceived.

First, how are we to understand the premise, "scientific success at the instrumental level"? One way is suggested in Fine's characterization of Vaihinger as holding that "fictions are *justifiable* to the extent to which they prove themselves useful in life's activities" (p. 7). This sounds like something that might have been written by John Dewey. And, indeed, inferences from practical success to "literal truth" are questionable. But moderate scientific realists like myself would emphasize success in *experiments* deliberately designed and executed in order rigorously to test specific hypotheses claiming a good fit between a model and some concrete system. So the premise is not just success in experience, but experimental success.

Second, the conclusion for a moderate realist is not "the literal truth of general principles" but merely the appropriate fit of specific models to some aspects of the world. In my framework, realist conclusions go only as high as the fit of specific models to designated aspects of the world. There is no inference to the literal truth of more general principles which serve mainly to define the most abstract models. These principles are true only of the models they define.[14] With these changes in both premises and conclusions, the inference in question is no longer obviously fallacious but arguably quite reasonable.

Finally, the view that scientific models are ontologically like works of fiction in being imaginary creations not only does not uniquely support fictionalism, but is compatible with a moderate realism.[15] There is nothing in this notion of a scientific model that prevents identifying elements of models with things traditionally classified as "unobservable." On the other hand, as discussed earlier in this chapter, some elements of models may not be identified with anything in the world.

7. CULTURAL CONSIDERATIONS

There are cultural dangers, particularly in the United States, in the general claim that scientific models are fictions. For religiously inspired antiscience movements, such as "creation science," learning that respected philosophers of science think that science is just a matter of fictions would be welcome news indeed. Even worse, for those who embrace Walton's (1990) account of fiction, scientific representation is a matter of "make-believe." We need only remember how creationists and other opponents of evolutionary theory welcomed Karl Popper's claims that evolutionary theory is not falsifiable and thus not a scientific theory but, rather, a metaphysical research program.[16] Niceties aside, this view was taken as providing justification for the metaphysical claims of religion as being on a par with those of science. Those of us who agree with Vaihinger that religions' claims about gods, prophets, and the like are fictional should, in the present cultural climate, be reluctant to place the claims of modern science on the same footing.[17] So the view that scientific claims are fictional is not one that supporters of a genuine scientific ethos should encourage. Indeed, in the present cultural climate, it would be irresponsible publicly to promulgate such views in these terms. At the very least, supporters of this view should come up with another description of their views. Perhaps "instrumentalism" or "pragmatism" would do. "Fictional" is far too loaded a word.

I must admit that even the view that, ontologically, scientific models, like works of fiction, are creations of human imagination raises some of the same problems, especially if, as is likely, the distinction between ontology and function is ignored. Nevertheless, this version of the cultural problem may be easier to counter than the blanket claim that scientific models are works of fiction. The idea, much beloved by scientists, that doing science involves creativity and imagination has long been accepted by the general population as a positive, indeed, humanizing, feature of science. Moreover, the idea that scientific models, unlike religious claims, are subject to rigorous experimental testing provides a means of demarcating science from religion. This, of course, sounds a bit like Popper's invocation of the now rejected distinction between "discovery" and "justification" in support of his views that doing science is a matter of conjectures and refutations, and that scientific claims must be falsifiable. One difference is that we now distinguish between claims that a particular model well fits its target, which can sometimes be clearly falsified, and larger units, paradigms, research programs, or perspectives, which are not subject to simple falsification but can, over time, be rejected. In addition, we no longer relegate the creative process to a mysterious "creativity" subject at most to psychological investigation. As shown by the work of Darden (1991), Nickles (1980), Nersessian, Morrison, and many others, the scientific imagination is guided by identifiable heuristic strategies and subject matter specific constraints.

Thus, the view that scientific models are initially a product of scientific imagination is indeed far easier to defend against misuse than a blanket claim that they are works of fiction.

NOTES

1. I wish to thank Mauricio Suárez for organizing the conference on fictions in science in Madrid, February 2006, and all the participants both for useful discussions and for providing everyone involved with drafts of their papers for this volume. I have been particularly influenced by the contributions of Margaret Morrison, Paul Teller, and Eric Winsberg.
2. I am grateful to Gabriele Contessa for helping me to appreciate the importance of distinguishing the ontology of models from their functions. See Contessa (in press).
3. I have earlier suggested that *War and Peace* can be regarded as an elaborate model of a possible world (Giere, 2006, chap. 4, note 6, p. 128).
4. The idea that everyday planning involves imagining models of possible scenarios appears already in the third edition of my elementary textbook on scientific reasoning (Giere, 1991, p. 27). That I have not earlier made this point in my more professional writings indicates that I thought it sufficiently unproblematic so as not to require professional mention. In light of recent discussions, I could not have been more mistaken.
5. The index to Blackwell's *Companion to Cognitive Science* (Bechtel & Graham, 1998), for example, contains no entries for either 'imagination' or 'planning.'
6. See, for example, Voltolini (2006) for an introduction to some of this literature.
7. Here one might recall June Goodfield's book *An Imagined Universe* (1981), which traces the work of an immunologist during the 1970s, and Howard Gruber's study of Darwin (1981).
8. Thompson-Jones (in press) considers, but in the end rejects, the claim that abstract models are mathematical objects. He does, however, claim we cannot ascribe temporal predicates to abstract models. Barberousse and Ludwig (this volume) claim that terms do not have the same meaning when applied to objects in a model as when applied to concrete objects.
9. I do, however, have some more to say (Giere, 2006, chap. 4; in press-a; in press-b).
10. This is one of the main lessons of Morgan and Morrison (1999).
11. For extended discussions of the related notions of abstraction and idealization, see Morrison (this volume) and Teller (this volume).
12. For an extended defense of this view based on an actual scientific text, see Held (this volume).
13. Unlike Vaihinger (and Fine), however, Scheffler, like most analytic philosophers of science of his time, was most concerned with the semantic issue of the meaning of theoretical terms.
14. This theme is developed in much more detail in Giere (2006, chap. 4).
15. This point was urged on me in private communication by Peter Godfrey-Smith.
16. Popper (1978) later repudiated these earlier views, but the damage was done.
17. In the present cultural climate, Vaihinger's distinction between the "genuine fictions" of religion and the "semi-fictions" of science is far too subtle.

References

Abraham, F., Broughton, J., Bernstein, N., and Kaxiras, E. (1998). Spanning the length scales in dynamic simulation. *Computers in Physics, 12*(6), 538–546.

Akerlof, G. A. (1970). The market for 'lemons': Quality uncertainty and the market mechanism. *Quarterly Journal of Economics, 84*, 488–500.

Alexandrova, A. (2006). Connecting economic models to the real world: Game theory and the FCC spectrum auctions. *Philosophy of the Social Sciences, 36*, 173–192.

Ankeny, R. A. (2000), Fashioning descriptive models in biology: Of worms and wiring diagrams. *Philosophy of Science, 67*, S260–S272.

Ankeny, R. A. (2001a). Model organisms as models: Understanding the "lingua franca" of the human genome project. *Philosophy of Science, 68*, S251–S261.

Ankeny, R. A. (2001b). The natural history of C. *elegans* research. *Nature Reviews Genetics, 2*, 474–478.

Batterman, R. (2005). Critical phenomena and breaking drops: Infinite idealizations in physics. *Studies in History and Philosophy of Modern Physics, 36*, 225–244.

Bechtel, W. (1993). Integrating cell biology by creating new disciplines: The case of cell biology. *Biology and Philosophy, 8*, 277–299.

Bechtel, W., & Graham, G. (Eds.). (1998). *A companion to cognitive science.* Oxford: Blackwell.

Black, M. (1962). *Models and metaphors.* Ithaca, NY: Cornell University Press.

Blumberg, A. E., & Feigl, H. (1931). Logical positivism: A new movement in European philosophy. *Journal of Philosophy, 28*, 1–96.

Bogen, J., & Woodward, J. (1988). Saving the phenomena. *Philosophical Review, 97*, 303–352.

Bohr, N. (1913). On the constitution of atoms and molecules. *Philosophical Magazine, 26*, 1–25, 476–502, 857–875.

Bohr, N. (1921). Atomic structure. *Nature, 107*, 1–11. Reprinted in *Niels Bohr collected works*, vol. 4, pp. 71–82. vol. 4: The Periodic System (1920–1923), J. R. Neilsen (ed.). Amsterdam: North-Holland Publishing, pp. 71–82.

Bokulich, A. (2008a). *Reexamining the quantum-classical relation: Beyond reductionism and pluralism.* Cambridge University Press.

Bokulich, A. (2008b). Can classical structures explain quantum phenomena? *British Journal for the Philosophy of Science, 59*(2), 217–235.

Bolker, J. A. (1995). Model systems in developmental Biology. *BioEssays, 17*: 451–455.

Bono, H., Ogata, H., Goto, S., et al. (1998). Reconstruction of amino acid biosynthesis pathways from the complete genome sequence. *Genome Research, 8*, 203–210.

Bono, J. (1990). Science, discourse and literature: The role/rule of metaphor in science. In S. Peterfreund (Ed.), *Literature and science: Theory and practice* (pp. 59–89). Boston: Northeastern University Press.

Boumans, M. (1999). Built-in justification. In M. S. Morgan & M. Morrison (Eds.), *Models as mediators: Perspectives on natural and social science* (pp. 66–96). Cambridge: Cambridge University Press.

Boumans, M. (2006). The difference between answering a 'why'-question and answering a 'how much'-question. In J. Lenhard, G. Küppers & T. Shinn (Eds.), *Simulation: Pragmatic Constructions of Reality. Sociology of the Sciences Yearbook* (pp. 107–124). New York: Springer.

Boumans, M., & Morgan, M. S. (2001). *Ceteris paribus* conditions: Materiality and the application of economic theories. *Journal of Economic Methodology, 8*, 11–26.

Brandom, R. (1994). *Making it explicit.* Cambridge, MA: Harvard University Press.

Brenner, S. (1988). Foreword. In W. B. Wood and the Community of *C. elegans* Researchers (Eds.), *The nematode* Caenorhabditis elegans (pp. ix–x). Cold Spring Harbor, NY: Cold Spring Harbor Laboratory Press.

Broughton, J., et al. (1999). Concurrent coupling of length scales: Methodology and application. *Physical Review B, 60*(4), 2391–2403

Buchwald, J., & Warwick, A. (Eds.). (2004). *Histories of the electron: The birth of microphysics.* Cambridge, MA: MIT Press.

Busch, P., Lahti, P., & Mittelstaedt, P. (1991). *The quantum theory of measurement.* Berlin: Springer-Verlag.

Caldwell, B. J. (1982). *Beyond positivism: Economic methodology in the twentieth century.* London: Allen & Unwin.

Callender, C. (2001) "Taking Thermodynamics Too Seriously" *Studies in History and Philosophy of Modern Physics, 32*: 539–53.

Campbell, J. (2000). *Artificial viscosity for multi-dimensional hydrodynamics codes.* http://cnls.lanl.gov/Highlights/2000–09/article.htm

Campbell, N. (1920). Atomic structure. *Nature, 106*(2665), 408–409.

Cangelosi, A., & Parisi, D. (1997). A neural network model of *Caenorhabditis elegans*: The circuit of touch sensitivity. *Neural Processing Letters, 6*, 91–98.

Caramana, E. J., Shashkov, M. J., & Whalen, P. P. (1988). Formulations of artificial viscosity for multi-dimensional shock wave computations. *Journal of Computational Physics, 144*, 70–97.

Carroll, S., Grenier, J., & Weatherbee, S. (2001). *From DNA to diversity.* Malden, MA: Blackwell.

Cartwright, N. (1983). *How the laws of physics lie.* Oxford: Oxford University Press.

Cartwright, N. (1989). *Nature's capacities and their measurement.* Oxford: Oxford University Press.

Cartwright, N. (1998). Capacities. In J. B. Davis, D. W. Hands, & U. Mäki (Eds.), *The handbook of economic methodology* (pp.). Cheltenham: Edgar Elgar.

Cartwright, N. (1999a). *The dappled world.* Cambridge: Cambridge University Press.

Cartwright, N. (1999b). *The vanity of rigour in economics: Theoretical models and Galilean experiments.* Discussion paper series 43/99. Centre for Philosophy of Natural and Social Science.

Chafe, W. L. (1970). *Meaning and the structure of language.* Chicago: University of Chicago Press.

Chalfie, M., Sulston, J. E., White, J. C., et al. (1985). The neural circuit for touch sensitivity in *Caenorhabditis elegans. Journal of Neuroscience, 5*, 959–964.

Chalmers, D. J. (1996). *The conscious mind. In search of a fundamental theory.* New York: Oxford University Press.

Chalmers, D. J. (2002). Does conceivability entail possibility? In T. Gendler & J. Hawthorne (Eds.), *Conceivability and possibility* (pp. 145–200). Oxford: Oxford University Press.

Chang, H. (2004). *Inventing temperature*. Oxford: Oxford University Press.
Chelikowski, J., & Ratner, M. (2001). Nanoscience, nanotechnology, and modeling. *Computing in Science and Engineering, 3*(4), 40–41.
Clarke, A. E., & Fujimura, J. (Eds.). (1992). *The right tools for the job: At work in twentieth-century life sciences*. Princeton, NJ: Princeton University Press.
Cohen, M. R. (1923). On the logic of fictions. *Journal of Philosophy, 20*, 477–488.
Contessa, G. (2007). Representation, interpretation, and surrogative reasoning. *Philosophy of Science, 71*, 48–68.
Contessa, G. (in press). Scientific models as fictional objects. *Synthese*.
Creager, A. N. H. (2002). *The life of a virus: Tobacco mosaic virus as an experimental model, 1930–1965*. Chicago: University of Chicago Press.
Currie, G. (1985). What is fiction? *Journal of Aesthetics and Art Criticism*, 385–392.
Currie, G. (1990). *The nature of fiction*. Cambridge University Press.
Currie, G., & Ravenscroft, I. (2002), *Recreative minds*, Clarendon Press.
Darden, L. (1991). *Theory change in science: Strategies from Mendelian genetics*. New York: Oxford University Press.
Darrigol, O. (1992). *From c-numbers to q-numbers: The classical analogy in the history of quantum theory*. Berkeley: University of California Press.
Davidson, D. (1984). On the very idea of a conceptual scheme. In *Inquiries into truth and interpretation* (pp. 183–198). Oxford: Oxford University Press.
Davidson, D. (1986). A nice derangement of epitaphs. In E. LePore (Ed.), *Truth and interpretation* (pp. 433–446). Oxford: Blackwell.
de Chadarevian, S. (1998). Of worms and programmes: *Caenorhabditis elegans* and the study of development. *Studies in the History and Philosophy of Science, 29*, 81–105.
de Chadarevian, S., & Hopwood, N. (2004). *Models: The third dimension of science*. Stanford, CA: Stanford University Press.
Delos, J. B., & Du, M.-L. (1988). Correspondence principles: The relationship between classical trajectories and quantum spectra. *IEEE Journal of Quantum Electronics, 24*, 1445–1452.
Delos, J. B., Knudson, S. K., & Noid, D. W. (1983). Highly excited states of a hydrogen atom in a strong magnetic field. *Physical Review A, 28*, 7–21.
Dewey, J. (1916). *Essays in experimental logic*. Chicago.
Doležel, L. (1998). *Heterocosmica: fiction and possible worlds*. Baltimore: Johns Hopkins University Press.
Du, M.-L., & Delos J. B. (1988). Effect of closed classical orbits on quantum spectra: ionization of atoms in a magnetic field. I. Physical picture and calculations. *Physical Review A, 38*, 1896–1912.
Duhem, P. (1977). *The aim and structure of physical theory*. New York: Atheneum.
Earman, J. (2003). *Spontaneous symmetry breaking for philosophers* (preprint).
Eco, U. (1979). *The role of the reader. Explorations in the semiotics of texts*. Bloomington: Indiana University Press.
Eco, U. (1983). *The name of the rose*. New York: G. K. Hall & Company.
Eco, U. (1990). *The limits of interpretation*. Bloomington: Indiana University Press.
Ehrenfest, P., & Ehrenfest, T. (1907). Über zwei bekannte Einwände gegen Boltzmann's H-Theorem. *Physische Zeitschrift, 8*, 311.
Elgin, C. Z. (1983). *With reference to reference*. Indianapolis, IN: Hackett.
Elgin, C. Z. (1991). Understanding in art and science. In P. French, T. Uehling, & H. Wettstein (Eds.), *Philosophy and the arts* (pp. 196–208). Notre Dame, IN: University of Notre Dame Press.
Elgin, C. Z. (1996). *Considered judgment*. Princeton, NJ: Princeton University Press.
Elgin, C. Z. (2004). True enough. *Philosophical Issues, 14*, 113–131.

Fine, A. (1970). Insolubility of the quantum measurement problem. *Physical Review D*, 2, 2783–2787.

Fine, A. (1986). Unnatural attitudes: Realist and instrumentalist attachments to science. *Mind*, 95, 149–79.

Fine, A. (1993). Fictionalism. *Midwest Studies in Philosophy, XVIII*, 1–18. Reprinted in this volume.

Fisher, R. A. (1918). The correlation between relatives on the supposition of Mendelian inheritance. *Transactions of the Royal Society of Edinburgh*, 52, 399–433.

Fisher, R. A. (1922). On the dominance ratio. *Proceedings of the Royal Society of Edinburgh*. 42, 321–341.

Fleck, L. (1979). *Genesis and development of a scientific fact*, F. Bradley and T. Trenn (Trans.). Chicago: University of Chicago Press.

Frank, J. (1970). *Law and the modern mind*. Gloucester, MA.

Frank, P. (1949). *Modern science and its philosophy*. Cambridge, MA.

Fraser, C. M., Gocayne, J. D., White, O., et al. (1995). The minimal gene complement of *Mycoplasma genitalium*. *Science*, 270, 397–403.

French, S. (2003). A model-theoretic account of representation (Or, I don't know much about art . . . but I know it involves isomorphism). *Philosophy of Science*, 70, 1472–1483.

French, S., & Ladyman, J. (1999). Reinflating the semantic approach. *International Studies in the Philosophy of Science*, 13, 103–121.

Friedman, M. (1953). The methodology of positive economics. In *Essays in positive economics* (pp.). Chicago: Chicago University Press.

Frigg, R. (2003). *Re-presenting scientific representation*. PhD dissertation, London School of Economics, London.

Frigg, R. (in press). Models and fiction. *Synthese* (2008).

Frigg, R., & Hartmann, S. (2006). Models in science. *The Stanford Encyclopedia of Philosophy*. http://plato.stanford.edu/archives/spr2006/entries/models-science/

Galison, P. (1987). *How experiments end*. Chicago: University of Chicago Press.

Galison, P. (1997). *Image and logic*. Chicago: University of Chicago Press.

Garton, W. R. S., & Tomkins, F. S. (1969). Diamagnetic Zeeman effect and magnetic configuration mixing in long spectral series of BA I. *Astrophysical Journal*, 158, 839–845.

Geiger, H., & Marsden, E. (1909). On a diffuse reflection of the α-particles. *Proceedings of the Royal Society*, 82, 495–500.

Gibson, D. G., Benders, G. A., Andrews-Pfannkoch C., et al. (2008). Complete chemical synthesis, assembly, and cloning of a mycoplasma genitalium genome. *Science Online*, 24 January.

Giere, R. N. (1988). *Explaining science: A cognitive approach*. Chicago and London: University of Chicago Press.

Giere, R. N. (1991). *Understanding scientific reasoning* (3rd ed.), New York: Harcourt Brace Jovanovich.

Giere, R. N. (1999). *Science without laws*. Chicago: Chicago University Press.

Giere, R. N. (2002). Scientific cognition as distributed cognition. In P. Carruthers, S. Stich, & M. Siegal (Eds.), *Cognitive bases of science* (pp.). Cambridge: Cambridge University Press.

Giere, R. N. (2004). How models are used to represent reality. *Philosophy of Science*, 71, 742–752.

Giere, R. N. (2006). *Scientific perspectivism*. Chicago: University of Chicago Press.

Giere, R. N. (in press-a). An agent-based conception of models and scientific representation. *Synthese*.

Giere, R. N. (in press-b). Models, metaphysics and methodology. In S. Hartmann, C. Hoefer, & L. Bovens (Eds.), *Nancy Cartwright's philosophy of science.* Oxford: Routledge.

Glass, J. I., Assad-Garcia, N., Alperovich, N., et al. (2006). Essential genes of a minimal bacterium. *Proceedings of the National Academy of Sciences USA, 103,* 425–430.

Godfrey-Smith, P. (2006). The strategy of model-based science. *Biology and Philosophy, 21,* 725–740.

Gombrich, H. (1961). *Art and illusion: A study in the psychology of pictorial representation.* Princeton, NJ: Princeton University Press.

Gombrich, H. (1984). *Meditations on a hobby horse.* Chicago University Press.

Goodfield, J. (1981). *An imagined world: A story of scientific discovery.* New York: Harper & Row.

Goodman, N. (1968). *Languages of art.* Indianapolis, IN: Hackett.

Gould, J. A. (1970). R. B. Perry on the origin of American and European pragmatism. *History of Philosophy, 8*(43), 1–50.

Granger, B. E. (2001). *Quantum and semiclassical scattering matrix theory for atomic photoabsorption in external fields.* PhD thesis, University of Colorado, Boulder.

Gruber, H. (1981). Darwin on man: A psychological study of scientific creativity. Chicago: University of Chicago Press.

Gutzwiller, M. (1971). Periodic orbits and classical quantization conditions. *Journal of Mathematical Physics, 12,* 343–358.

Gutzwiller, M. (1990). *Chaos in classical and quantum mechanics.* New York: Springer-Verlag.

Hacking, I. (1983). *Representing and intervening.* Cambridge: Cambridge University Press.

Hacking, I. (1984). Language, truth and reason. In M. Hollis & S. Lukes (Eds.), *Rationality and relativism* (pp. 48–66). Cambridge, MA: MIT Press.

Hacking, I. (1986). Making up people. In T. C. Heller, M. Sosna, & D. E. Wellbery (Eds.), *Reconstructing individualism* (pp. 222–236). Stanford, CA: Stanford University Press.

Hacking, I. (1992). The self-vindication of the laboratory sciences. In A. Pickering (Ed.), *Science as practice and culture* (pp. 29–64). Chicago: University of Chicago Press.

Hacking, I. (1999). *The social construction of what?* Cambridge, MA: Harvard University Press.

Hacking, I. (2002). *Historical ontology.* Cambridge, MA: Harvard University Press.

Hagberg, G. L. (1994). *Meaning and interpretation: Wittgenstein, Henry James, and literary knowledge.* Ithaca, NY: Cornell University Press.

Haggerty, M. R., Spellmeyer, N., Kleppner, D., & Delos, J. B. (1998). Extracting classical trajectories from atomic spectra. *Physical Review Letters, 81,* 1592–1595.

Hanna, P., & Harrison, B. (2004). *Word and world.* Cambridge: Cambridge University Press.

Harris, T. (2003). Data models and the acquisition and manipulation of data. *Philosophy of Science, 70,* 1508–1517.

Haugeland, J. (1998). *Having thought.* Cambridge, MA: Harvard University Press.

Hausman, D. (1990). Supply and demand explanations and their *ceteris paribus* clauses. *Review of Political Economy, 2,* 168–187.

Hausman, D. (1992). *The inexact and separate science of economics.* Cambridge: Cambridge University Press.

Hegselmann, R., & Siegwart, G. (1991). Zur Geschichte der 'Erkenntnis'. *Erkenntnis, 35*, 461–471.

Heidegger, M. (1950). *Holzwege*. Frankfurt am Main: Vittorio Klostermann. English tr. 2002. *Off the beaten track* (J. Young, Trans.). Cambridge: Cambridge University Press.

Heilbron, J. (1967). The Kossel-Sommerfeld theory and the ring atom. *Isis, 58*(4), 450–485.

Hempel, C. G. (1965). *Aspects of scientific explanation and other essays in the philosophy of science*. New York: Free Press.

Hempel, C. G. (1975). The old and the new 'Erkenntnis'. *Erkenntnis, 9*, 1 4 .

Hesse, M. (1963). *Models and analogies in science*. Notre Dame, IN: University of Notre Dame Press.

Hobbes, T. (1985). Dialogus physicus (S. Schaffer, Trans.). In S. Shapin & S. Schaffer, *Leviathan and the air pump* (pp. 346–391). Princeton, NJ: Princeton University Press.

Holle, A., Main, J., Wiebusch, G., Rottke, H., & Welge, K. H. (1988). Quasi-Landau spectrum of the chaotic diamagnetic hydrogen atom. *Physical Review Letters, 61*, 161–164.

Horwich, P. (1991). On the nature and norms of theoretical commitment. *Philosophy of Science, 58*, 1–14.

Hughes, R. I. G. (1989). Bell's theorem, ideology, and structural explanation. In *Philosophical consequences of quantum theory: Reflections on Bell's theorem* (pp. 195–207). Notre Dame, IN: University of Notre Dame Press.

Hughes, R. I. G. (1997). Models and representation. *Philosophy of Science, 64*, S325–S336.

Hughes, R. I. G. (1999). The Ising model, computer simulation, and universal physics. In M. Morrison & M. Morgan (Eds.), *Models as mediators* (pp. 97–145). Cambridge: Cambridge University Press.

Humphreys, P. (1994). Numerical experimentation. In P. Humphreys (Ed.), *Patrick Suppes, Scientific philosopher* (Vol. 2, pp.). Dordrecht: Kluwer.

Humphreys, P. (2002). Computational models. *Philosophy of Science, 69*, S1–S11.

Humphreys, P. (2004). *Extending ourselves: Computational science, empiricism, and scientific method*. New York, Oxford University Press.

Hutchison, T. W. (1938). *The significance and basic postulates of economic theory*. London: MacMillan.

John, E. (1998). Reading fiction and conceptual knowledge: Philosophical thought in literary context. *Journal of Aesthetics and Art Criticism, 56*, 331–348.

Jones, M., & Cartwright, N. (2004). *Correcting the model: Idealisation and abstraction in the sciences*. Rodopi.

Kalderon, M. E. (2005). *Fictionalism in metaphysics*. Oxford University Press.

Kitano, H., Amáis, S., & Luke, S. (1988). The perfect *C. elegans* project: An initial report. *Artificial Life, 4*, 141–156.

Kitano, H., S. Hamahashi, and S. Luke (1988), "The Perfect C. elegans Project: An Initial Report", Artificial Life 4: 141–156.

Kitcher, P. (1993). *The advancement of science*. Oxford: Oxford University Press.

Klein, U. (2003). *Experiments, models, paper tools*. Stanford, CA: Stanford University Press.

Kleppner, D., & Delos, J. B. (2001). Beyond quantum mechanics: Insights from the work of Martin Gutzwiller. *Foundations of Physics, 31*, 593–612.

Knuuttila, T. (2005). Models, representation, and mediation. *Philosophy of Science, 72*, 1260–1271.

Knuuttila, T. (2006). From representation to production: Parsers and parsing in language technology. In J. Lenhard, G. Küppers, & T. Shinn (Eds.), *Simulation: Pragmatic constructions of reality. Sociology of the Sciences Yearbook* (pp. 41–55). New York: Springer.

Knuuttila, T., & Voutilainen, A. (2003). A parser as an epistemic artefact: A material view on models. *Philosophy of Science, 70*, 1484–1495.

Kohler, R. (1994). *Lords of the fly. Drosophila genetics and the experimental life.* Chicago: University of Chicago Press.

Kuhn, T. (1977). A function for thought experiments. In *The essential tension* (pp. 240–265). Chicago: University of Chicago Press.

Kuorikoski, J., Lehtinen, A., & Marchionni, C. (2007). *Economics as robustness analysis.* Paper presented at Models and Simulations 2, Tilburg, The Netherlands. http://philsci-archive.pitt.edu/archive/00003550/

Lambrecht, K. (1994). *Information structure and sentence form: Topic, focus and the mental representations of discourse referents.* Cambridge; Cambridge University Press.

Lehtinen, A., & Kuorikoski, J. (2007). Computing the perfect model: Why do economists shun simulation. *Philosophy of Science, 74*, 304–329.

Lenhard, J. (2006). Surprised by a nanowire: Simulation, control, and understanding. *Philosophy of Science, 73*, 605–616.

Lester, E. (1966, September 18). Playwright Edward Albee: 'I am still in process.' *The New York Times*, Arts & Leisure, p. 113.

Levins, R. (1966). The strategy of model-building in population biology. *American Scientist, 54*, 421–431.

Lewis, D. (1979). Attitudes *de dicto* and *de re. Philosophical Review, 88*, 513–543.

Lewis, D. (1982). Logic for equivocators. *Noûs, 16*, 431–441. Reprinted in Lewis, D. (1983). *Philosophical papers.* Oxford: Oxford University Press.

Lewis, D. (1983). Truth in fiction. In D. Lewis (Ed.), *Philosophical papers* (Vol. 1, pp. 261–280). Oxford: Oxford University Press. (Original work published 1978)

Lipton, P. (1991). *Inference to the best explanation.* Routledge.

Lopes, D. (1996). *Understanding pictures.* Oxford: Oxford University Press.

Lu, K. T., Tomkins, F. S., & Garton, W. R. S. (1978). Configuration interaction effect on diamagnetic phenomena in atoms: Strong mixing and Landau regions. *Proceedings of the Royal Society of London A, 362*, 421–424.

Lucas, R. E., Jr. (1980). Methods and problems in business cycle theory. *Journal of Money, Credit and Banking, 12*(4), 696–715.

Lucas, R. E., Jr. (1981). *Studies in business-cycle theory.* Cambridge, MA: MIT Press.

Lycan, W. (1994). *Modality and meaning.* Dordrecht: Kluwer Academic.

Lynch, M. (1990). The externalized retina: Selection and mathematization in the visual documentation of objects in life sciences. In M. Lynch & S. Woolgar (Eds.), *Representation in scientific practice* (pp. 153–186). Cambridge, MA: MIT Press.

Machlup, F. (1955). The problem of verification in economics. *Southern Economic Journal, 22*, 1–21.

Machlup, F. (1967). Theories of the firm: Marginalist, behavioral, managerial. *American Economic Review, LVII*, 1–33.

Machlup, F. (1978). *Methodology of economics and other social sciences.* New York: Academic Press.

Magnani, L. (1999). *Model-based reasoning in scientific discovery.* Dordrecht: Kluwer Academic.

Main, J., Weibusch, G., Holle, A., & Welge, K. H. (1986). New quasi-Landau structure of highly excited atoms: The hydrogen atom. *Physical Review Letters, 57*, 2789–2792.

Mäki, U. (1990). *Studies in realism and explanation in economics.* Helsinki: Suomalainen tiedeakatemia.

Mäki, U. (1992). On the method of isolation in economics. *Poznań Studies in the Philosophy of Science and Humanities, 26*, 316–351.

Mäki, U. (1994). Reorienting the assumptions issue. In R. Backhouse (Ed.), *New directions in economic methodology* (pp. 236–256). London: Routledge.

Mäki, U. (1996). Scientific realism and some peculiarities of economics. In R. S. Cohen, R. Hilpinen, & Q. Renzong (Eds.), *Realism and anti-realism in the philosophy of science* (pp. 427–447). Dordrecht: Kluwer.

Mäki, U. (2002). *Fact and fiction in economics.* Cambridge: Cambridge University Press.

Mäki, U. (2004). Realism and the nature of theory: A lesson from J. H. von Thünen for economists and geographers. *Environment and Planning A, 36,* 1719–1736.

Mäki, U. (2005). Models are experiments, experiments are models. *Journal of Economic Methodology, 12,* 303–315.

Mäki, U. (2006). Remarks on models and their truth. *Storia del Pensiero Economico, 3,* 7–19.

Mäki, U. (in press), Realistic realism about unrealistic models, to appear in H. Kincaid and D. Ross (Eds.), *The Oxford Handbook of the Philosophy of Economics.* Oxford University Press.

Matzkin, A. (2006). *Can Bohmian trajectories account for quantum recurrences having classical periodicities?* arXiv.orgpreprint quant-ph/0607095.

Maxwell, J. C. (1965). *The scientific papers of James Clerk Maxwell* (W. D. Niven, Ed.). New York: Dover.

McDowell, J. (1994). *Mind and world.* Cambridge, MA: Harvard University Press.

McKeon, R. (Ed.). (1973). *Introduction to Aristotle.* Chicago: University of Chicago Press.

McMullin, E. (1985). Galilean idealization. *Studies in History and Philosophy of Science, 16,* 247–273.

Morgan, M. S. (2003). Experiments without material intervention: Model experiments, virtual experiments, and virtually experiments. In H. Radder (Ed.), *Philosophy of scientific experimentation* (pp. 216–235). Pittsburgh: University of Pittsburgh Press.

Morgan, M. S. (2005). Experiments versus models: New phenomena, inference and surprise. *Journal of Economic Methodology, 12,* 317–329.

Morgan, M. S., & Morrison, M. (Eds.). (1999). *Models as mediators. Perspectives on natural and social science.* Cambridge: Cambridge University Press.

Morrison, M. (1998). Modelling nature: Between physics and the physical world, *Philosophia Naturalis, 35,* 65–85.

Morrison, M. (1999). Models as autonomous agents. In M. S. Morgan & M. Morrison (Eds.), *Models as mediators: Perspectives on natural and social science* (pp. 38–65). Cambridge: Cambridge University Press.

Morrison, M. (2000). *Unifying scientific theories: Physical concepts and mathematical structures.* Cambridge: Cambridge University Press.

Morrison, M. (2002). Modelling populations: Pearson and Fisher on Mendelism and biometry. *British Journal for the Philosophy of Science, 53,* 39–68.

Morrison, M. & Morgan, M.S. (1999). Models as mediating instruments. In M. S. Morgan and M. Morrison (Eds.), *Models as mediators. Perspectives on natural and social science* (pp. 10–37). Cambridge: Cambridge University Press.

Musgrave, A. (1981). 'Unreal assumptions' in economic theory: The F-Twist untwisted. *Kyklos, 34,* 377–387.

Musil, R. (1906). *Young Törless* (E. Wilkins & E. Kaiser, Trans.). New York.

Nakano, A., et al. (2001). Multiscale simulation of nano-systems. *Computing in Science and Engineering. 3*(4), 56–66.

Nanney, D. L. (1983). The cytoplasm and the ciliates. *Journal of Heredity, 74,* 163–170.

Nersessian, N. J. (1992). How do scientists think? In R. N. Giere (Ed.), *Cognitive models of science* (pp. 3–44). Minnesota Studies in the Philosophy of Science (Vol. XV). Minneapolis: University of Minnesota Press.

Neurath, O. (1983). Encyclopedia as model. In R. S. Cohen & M. Neurath (Eds.), *Neurath: Philosophical papers* (pp. 145–158). Dordrecht: Reidel.

Nickles, T. (Ed). (1980). *Scientific discovery: Case studies.* Dordrecht: Reidel.

Norton, S., & Suppe, F. (2001). Why atmospheric modeling is good science. In C. Miller & P. Edwards (Eds.), *Changing the atmosphere* (pp. 68–105). Cambridge, MA: MIT Press.

Novitz, D. (1987). *Knowledge, fiction and imagination.* Philadelphia, PA: Temple University Press.

Nussbaum, M. (1990). *Love's knowledge.* New York: Oxford University Press.

Oja, M. F. (1988). Fictional history and historical fiction: Solzhenitsyn and Kiš as exemplars. *History and Theory, 27,* 111–124.

Papineau, D. (1988). Mathematical fictionalism. *International Studies in the Philosophy of Science, 2,* 151–175.

Pavel, T. G. (1986). *Fictional worlds.* Cambridge, MA: Harvard University Press.

Pearson, K. (1904). Mathematical contributions to the theory of evolution. XII. On a Generalized Theory of Alternative Inheritance with Special Reference to Mendel's Laws. *Proceedings of the Royal Society A, 203,* 53–86.

Popper, K. R. (1985). Natural selection and its scientific status. Reprinted with omissions In D. Miller (Ed.), *Popper selections* (pp.). Princeton, NJ: Princeton University Press. (Original work published 1978)

Portides, D. (2005). "Scientific Models and the Semantic View of Theories" *Philosophy of Science, 72:* 1287–1298.

Prialnik, D. (2000). *An Introduction to the Theory of Stellar Structure and Evolution.* Cambridge: Cambridge University Press.

Prince, E. F. (1981). Toward a taxonomy of given-new information. In P. Cole (Ed.), *Radical pragmatics* (pp. 233–237). New York: Academic Press.

Putnam, H. (1978). Literature, science, and reflection. In *Meaning and the moral sciences* (pp.). London: Routledge and Kegan Paul.

Quine, W. v. O. (1953). Two dogmas of empiricism. In *From a logical point of view* (pp. 139–160). Cambridge, MA: Harvard University Press.

Quine, W. v. O. (1969). Propositional objects. In *Ontological relativity and other essays* (pp. 20–46). New York: Columbia University Press.

Rader, K. A. (2004). *Making mice: Standardizing animals for American biomedical research, 1900–1955.* Princeton, NJ: Princeton University Press.

Railton, P. (1980). *Explaining explanation: A realist account of scientific explanation and understanding.* PhD dissertation, Princeton University, Princeton, NJ.

Ramsey, F. P. (1990). General propositions and causality. In *Philosophical papers* (pp.). Cambridge: Cambridge University Press. (Original work published 1929)

Redhead, M. (1980). Models in physics. *British Journal for the Philosophy of Science, 31,* 145–163.

Rheinberger, H. J. (1995). From microsomes to ribosomes: 'Strategies' of 'representation', 1935–1955. *Journal of the History of Biology, 48,* 49–89.

Rheinberger, H. J. (1997). *Toward a history of epistemic things.* Stanford, CA: Stanford University Press.

Robbins, L. (1935). *An essay on the nature and the significance of economic science* (2nd ed.). London: MacMillan.

Ronen, R. (1994). *Possible worlds in literary theory.* Cambridge: Cambridge University Press.

Rorty, R. (1980). *Philosophy and the mirror of nature.* Oxford: Basil Blackwell.

Rosen, G. (1994). What is constructive empiricism? *Philosophical Studies, 74,* 143–178.

Rosen, G. (2005). Problems in the history of fictionalism. In M. E. Kalderon (Ed.), *Fictionalism in Metaphysics* (pp. 14–64).

Rouse, J. (1987). *Knowledge and power.* Ithaca, NY: Cornell University Press.

Rouse, J. (2002). *How scientific practices matter.* Chicago: University of Chicago Press.

Rudd, R. E., and Broughton, J. Q (2000). Concurrent coupling of length scales in solid state systems. *Physica Status Solidi B, 217,* 251–291.

Russell, B. (1968). On denoting. In *Logic and knowledge* (pp. 39–56). New York: Capricorn.

Rutherford, E. (1911). The scattering of α and β particles by matter and the structure of the atom. *Philosophical Magazine, 21,* 669–688.

Ryckman, T. (1991). Conditio sine qua non: 'Zuordnung' in the early epistemologies of Cassirer and Schlick. *Synthèse, 88,* 57–95.

Sainsbury, M. (2005). *Reference without referents.* Oxford: Oxford University Press.

Salmon, W. (1984). *Scientific explanation and the causal structure of the world.* Princeton, NJ: Princeton University Press.

Sapp, J. (1987), *Beyond the gene.* Oxford: Oxford University Press.

Scanlon, T. M (1998). *What we owe to each other.* Cambridge, MA: Harvard University Press.

Schaffner, K.F. (2000). Behavior at the organismal and molecular levels: The case of *C. elegans. Philosophy of Science, 67,* S273–S288.

Schaper, E. (1965). The Kantian 'as if' and its relevance for aesthetics. *Proceedings of the Aristotelian Society, 65,* 2, 19–34.

Schaper, E. (1966). The Kantian thing-in-itself as a philosophical fiction. *Philosophical Quarterly, 16,* 233–243.

Scheffler, I. (1963). *The anatomy of inquiry: Philosophical studies in the theory of science.* New York: Knopf.

Scheffler, I. (1982). Epistemology of objectivity. In *Science and subjectivity* (pp. 114–124). Indianapolis, IN: Hackett.

Schelling, T.C. (1978). *Micromotives and macrobehavior.* New York: Norton.

Schlick, M. (1932). Positivism and realism. In *Logical positivism* (A. J. Ayer, Ed., pp. 82–107). New York: . (Original work published 1932)

Seigel, D. (1991). *Innovation in Maxwell's electromagnetic theory.* Cambridge: Cambridge University Press.

Sejnowski, T. J., Koch, C., & Churchland, P. S. (1988). Computational neuroscience. *Science, 241,* 1299–1306.

Sklar, L. (1993). Idealization and explanation: A case study from statistical mechanics. In *Minnesota studies in philosophy, Vol. XVII, Philosophy of science* (pp. 258–270.). Notre Dame, IN: University of Notre-Dame Press.

Smith, E. E., & Medin, D. L. (1981). *Categories and concepts.* Cambridge, MA: Harvard University Press.

Smith, G. E. (2004). J. J. Thomson and the electron, 1897–1899. In J. Buchwald & A. Warwick (Eds.), *Histories of the electron: The birth of microphysics* (pp. 21–76). Cambridge, MA: MIT Press.

Solow, R. M. (1997). How did economics get that way and what way did it get? *Daedalus, 126,* 39–58.

Spariosu, M. (1989). *Dionysus reborn: Play and the aesthetic dimension in modem philosophical and scientific discourse.* Ithaca, NY: .

Stanford, P. K. (2006). *Exceeding our grasp: Science, history and the problem of unconceived alternatives.* Oxford: Oxford University Press.

Suárez, M. (1999). Theories, models, and representations. In L. Magnani, N. J. Nersessian, & P. Thagard (Eds.), *Model-based reasoning in scientific discovery* (pp. 75–83). New York: Kluwer.

Suárez, M. (2003). Scientific representation: Against similarity and isomorphism. *International Studies in the Philosophy of Science, 17,* 225–244.

Suárez, M. (2004a). Quantum selections, propensities and the problem of measurement. *British Journal of Philosophy of Science, 55*(2), 219–255.

Suárez, M. (2004b). An inferential conception of scientific representation. *Philosophy of Science, 71*, 767–779.

Sugden, R. (2002). Credible worlds: The status of the theoretical models in economics. In U. Mäki (Ed.), *Fact and fiction in economics: Models, realism, and social construction* (pp. 107–136). Cambridge: Cambridge University Press.

Suppes, P. (1962). Models of data. In E. Nagel, P. Suppes, & A. Tarski (Eds.), *Logic, methodology and philosophy of science: Proceedings of the 1960 international congress* (p. 252–261). Stanford, CA: Stanford University Press.

Suppes, P. (1974). The structure of theories and the analysis of data. In F. Suppe (Ed.), *The structure of scientific theories* (pp. 266–283). Urbana, IL University of Illinois Press.

Swoyer, C. (1991). Structural representation and surrogative reasoning. *Synthese, 87*, 449–508.

Tayler, R. J. (1970) *The Stars: their structure and evolution.* Cambridge: Cambridge University Press.

Teller, P. (2001). Twilight of the perfect model model. *Erkenntnis, 55*, 393–415.

Teller, P. (2008). Representation in science. In M. Curd & S. Psillos (Eds.), *The Routledge companion to the philosophy of science.* pp. 435–441.

Teller, P. (in preparation). Some dirty little secrets about truth.

Thomson, J. J. (1903). *Conduction of electricity through gases.* Cambridge University Press.

Thomson-Jones, M. (in press). Missing systems and the face value practice. *Synthese.*

Tufte, E. R. (1997). *Visual and statistical thinking: Displays of evidence for making decisions.* Cheshire, CT: Graphics Press.

Uebel, T. (1992). *Overcoming logical positivism from within.* Amsterdam: Rodopi.

Vaihinger, H. (1911). *Die Philosophie des Als Ob.* Berlin: Verlag von Reuther & Reichard.

Vaihinger, H. (1924). *The philosophy of 'as if'* (C. K. Ogden, Trans.). London: Kegan Paul. (Original work published 1911)

Vaihinger, H. (1952). *The philosophy of 'as if': A system of the theoretical, practical and religious fictions of mankind* (C. K. Ogden, Trans.). London: Lund Humphries. (Original work published 1911)

Van Fraassen, B. (1980). *The scientific image.* Oxford: Oxford University Press.

Van Fraassen, B. (1985). *Laws and symmetry.* Clarendon Press.

Vichniac, G. (1984). Simulating physics with cellular automata. *Physica D, 10*, 96–116.

Voltolini, A. (2006). *How ficta follow fiction: A syncretistic account of fictional entities.* New York: Springer.

Von Baeyer, H. C. (1995). The philosopher's atom. *Discover, 16*(11).

Von Neumann, J., & Richtmyer, R. D. (1950). A method for the numerical calculation of hydrodynamical shocks. *Journal of Applied Physics, 21*, 232–247.

Wade, N. (1999, December 14). Life is pared to basics; Complex issues arise. *The New York Times*, Science, p. 1.

Walsh, D. (1969). *Literature and knowledge.* Middletown, CT: Wesleyan University Press.

Walton, K. (1990). *Mimesis as make-believe: On the foundations of the representational arts.* Cambridge, MA: Harvard University Press.

Warwick, A. (1995). The sturdy protestants of science: Larmor, Trouton and the earth's motion through the ether. In J. Buchwald (Ed.), *Scientific practice* (pp.). Chicago: University of Chicago Press.

Warwick, A. (2003). *Masters of theory; Cambridge and the rise of mathematical physics.* Chicago: University of Chicago Press.

Weber, M. (2007). Redesigning the fruit fly: The molecularization of *Drosophila*. In A. Creager, E. Lunbeck, & M. N. Wise (Eds.), *Science without laws* (pp. 23–45). Durham, NC: Duke University Press.

Weinberg, S. (1995). *The quantum theory of fields, Vol. 1, Foundations*. Cambridge: Cambridge University Press.

Weisberg, M. (2006). Robustness analysis. *Philosophy of Science, 73,* 730–742.

Weisberg, M. (2007). Who is a modeler. *British Journal for the Philosophy of Science, 58,* 207–233.

Welch, G. R., Kash, M. M., Iu, C., Hsu, L., & Kleppner, D. (1989). Positive-energy structure of the diamagnetic Rydberg spectrum. *Physical Review Letters, 62,* 1975–1978.

White, H. (1984). The question of narrative in contemporary historical theory. *History and Theory, 23,* 1–33.

White, J. G., Southgate, E., Nicol Thomson, J., et al. (1986). The structure of the nervous system of the nematode *Caenorhabditis elegans:* The mind of a worm. *Philosophical Transactions of the Royal Society of London, Series B, 314,* 1–340.

Wilke, C. O., & Adami, C. (2002). The biology of digital organisms. *Trends in Ecology & Evolution, 17,* 528–532.

Wilson, C. (1983). Literature and knowledge. *Philosophy, 48,* 489–496.

Wilson, M. (2005). *Wandering significance*. Oxford: Oxford University Press.

Wimsatt, W. C. (1987). False models as means for truer theories. In M. H. Nitecki & A. Hoffman (Eds.), *Neutral models in biology* (pp. 23–55). New York: Oxford University Press.

Winsberg, E. (2003). Simulated experiments: Methodology for a virtual world. *Philosophy of Science, 70,* 105–125.

Winsberg, E. (2006a). Models of success vs. the success of models: Reliability without truth. *Synthese, 152,* 1–19.

Winsberg, E. (2006b). Handshaking your way to the top: Simulation at the nanoscale. *Philosophy of Science, 73,* 582–594.

Wollheim, R. (1980). *Art and its objects*. Cambridge University Press.

Woodward, J. (2003). *Making things happen: A theory of causal explanation*. Oxford: Oxford University Press.

Yablo, S. (1993). Is conceivability a guide to possibility? *Philosophy and Phenomenological Research, 53,* 1–42.

Contributors

Rachel A. Ankeny is an Associate Professor in History at the University of Adelaide. Previously she was Senior Lecturer in the Unit for History and Philosophy of Science and the Centre for Values, Ethics, and the Law in Medicine at the University of Sydney. Her current research focuses on epistemological issues in the contemporary life sciences and medicine, particularly modeling and case-based reasoning, and bioethics policy-making in contentious domains.

Anouk Barberousse is a research fellow at the Institut d'Histoire et de Philosophie des Sciences et des Techniques, Paris. She has worked on the history and philosophy of statistical mechanics, focusing on the diversity of its models. Her current research is devoted to the cognitive basis of scientific pictures and the epistemology of scientific simulation.

Alisa Bokulich is an Associate Professor in the Philosophy Department at Boston University. Her primary areas of research are the history and philosophy of physics, especially classical and quantum mechanics, and broader issues in the philosophy of science, such as scientific explanation and intertheoretic relations. She is the author of *Reexamining the Quantum-Classical Relation: Beyond Reductionism and Pluralism* (Cambridge University Press, 2008), and co-editor (with Peter Bokulich) of *Scientific Structuralism* (Boston Studies in the Philosophy of Science, Springer, in press).

Catherine Z. Elgin is Professor of the Philosophy of Education at the Graduate School of Education, Harvard University. She is the author of *Considered Judgement* (Princeton University Press, 1996), *Between the Absolute and the Arbitrary* (Cornell University Press, 1997), *With Reference to Reference* (Hackett Publishing Co., 1983), and co-author, with Nelson Goodman, of *Reconceptions in Philosophy and Other Arts and Sciences* (Hackett, 1988).

Arthur Fine is Professor of Philosophy, and Adjunct Professor of Physics and of History, at the University of Washington. His research concentrates on foundations of quantum physics and on interpretive issues relating to the natural and social sciences. His works include *The Shaky Game: Einstein, Realism and the Quantum Theory* (University of Chicago Press).

Ronald N. Giere is Professor of Philosophy Emeritus at the University of Minnesota as well as former director and continuing member of the Minnesota Centre for Philosophy of Science. Prof. Giere is a past president of the Philosophy of Science Assocation and a member of the editorial board of the journal *Philosophy of Science*. He is the author of *Understanding Scientific Reasoning* (5th ed., 2006), *Explaining Science: A Cognitive Approach* (1988), *Science Without Laws* (1999), and *Scientific Perspectivism* (2006). His current research focuses on the perspectival nature of scientific knowledge and on relations among naturalism, secularism, and liberal democracy.

Carsten Held is Professor of Philosophy of Science at the Universität Erfurt. His research concentrates on problems at the interface of philosophy of science and philosophy of language. His works include *Die Bohr-Einstein-Debatte* (Schöningh, 1998) and *Frege und das Grundproblem der Semantik* (Mentis, 2005)

Tarja Knuuttila is Research Fellow in the Department of Philosophy, University of Helsinki, Finland. She also teaches philosophy, semiotics, and science and technology studies at the University of Helsinki. Her main research interests include scientific representation and modeling with special emphasis on studying the various roles of techonological artifacts in science.

Pascal Ludwig is a senior lecturer in philosophy at University Paris-Sorbonne and a researcher at Institut Jean Nicod. He is currently working on the theory of reference in philosophy of language and on qualia, introspection and phenomenal concepts in philosophy of mind. He is a co-editor of Art and Mind, a special issue of The Monist (86: 4, 2003).

Margaret Morrison is Professor of Philosophy at the University of Toronto. Her work focuses on general philosophy of science, philosophy of physics, and the history of physics. She has authored the book *Unifying Scientific Theories: Physical Concepts and Mathematical Structures* (Cambridge University Press, 2000) and edited the book *Models as Mediators* (Cambridge University Press, 1999), with Mary Morgan. Some selected papers are "Spontaneous Symmetry Breaking: Theoretical Arguments and Philosophical Problems," in *Philosophical Aspects of Symmetries*, K. Brading and E. Castellani (Eds.) (Cambridge: Cambridge University Press, 2003); and "Modelling Populations: Pearson and Fisher on Mendelism and Biometry," *British Journal for the Philosophy of Science* (2002), 53, 39–68.

Joseph Rouse is the Hedding Professor of Moral Science in the Department of Philosophy, and Chair of the Science in Society Program at Wesleyan University. He is the author of *How Scientific Practices Matter* (Chicago, 2002), *Engaging Science* (Cornell, 1996), and *Knowledge and Power* (Cornell, 1987).

Mauricio Suárez is Associate Professor in Logic and Philosophy of Science at Complutense University. He previously held positions at Oxford, St. Andrews, Bristol, and Northwestern Universities. His main research interests lie in the philosophical foundations of physics (particularly quantum mechanics) and general epistemology of science, and he has published widely in both areas.

Paul Teller is a Professor in the Department of Philosophy at the University of California at Davis, and author of *An Interpretive Introduction to Quantum Field Theory* (Princeton University Press, 1995). He has published on many topics in philosophy of physics, as well as epistemology, metaphysics, and the philosophy of language. His current research focuses on what we can learn from the practice of science about the limitations on human knowledge and the repercussions for how we think about the human representational and epistemic enterprise.

Eric Winsberg is Associate Professor of Philosophy at the University or South Florida. His current research focuses on the role of models and simulations in the physical sciences, and on the foundations of statistical mechanics.

Index